HVAC Bible for Beginners

The Ultimate Guide to Advanced and Practical Problem Solving, Repair, Installation, and Maintenance of Residential and Commercial Climate Control [10 Books in 1]

© 2025 Bryan Allen Lawson - All rights reserved.

No part of this book may be reproduced, stored in a retrieval system, or transmitted in any form or by any means—electronic, mechanical, photocopying, recording, or otherwise—without prior written permission from the author or publisher, except as permitted by copyright law.

This book is protected by United States copyright laws and international treaties. It is intended solely for the personal, non-commercial use of the reader. Unauthorized distribution, sharing, or sale of copies of this book, in any format, is strictly prohibited and subject to legal penalties.

Every effort has been made to ensure that the information contained in this book is accurate and up-to-date at the time of publication. However, the author and publisher assume no responsibility for any errors, omissions, or changes in industry standards, safety regulations, or best practices that may occur over time. Readers are encouraged to verify the latest codes, laws, and guidelines from official sources before undertaking any HVAC-related work.

All trademarks, company names, and images mentioned in this book are the property of their respective owners. The use of such trademarks or names is solely for informational purposes and does not imply any affiliation, sponsorship, or endorsement by the mentioned entities.

Table of Contents

BOOK 1 INTRODUCTION TO HVAC: WHAT YOU NEED TO KNOW BEFORE GETTING STARTED 9

1.1 WHAT IS AN HVAC SYSTEM AND WHY IS IT ESSENTIAL? 10
- The Three Pillars of an HVAC System 10
- Why Is an HVAC System Essential? 11

1.2 TYPES OF HVAC SYSTEMS: RESIDENTIAL VS. COMMERCIAL – KEY DIFFERENCES AND WHY THEY MATTER 12
- Key Differences Between Residential and Commercial HVAC Systems 12
- 1. System Capacity and Power 12
- 2. System Structure and Configuration 13
- 3. Ventilation and Air Distribution 13
- 4. Maintenance Complexity and Repairs 14
- 5. Energy Efficiency and Automation 14
- Why Is It Important to Know These Differences? 14

1.3 HOW DOES AN HVAC SYSTEM WORK? – THE THERMODYNAMIC CYCLE EXPLAINED SIMPLY 15
- The 4 Phases of the Thermodynamic Cycle 15
- Compression: The Refrigerant is Compressed and Heats Up 16
- Condensation: Heat is Released into the Outdoor Environment 16
- Expansion: The Refrigerant Cools Down Rapidly 16
- Evaporation: The Indoor Air is Cooled 16
- And What About Heating? The Cycle Works in Reverse! 17
- The Role of Diagnostic Tools in HVAC Maintenance 17

1.4 ESSENTIAL TOOLS AND EQUIPMENT FOR WORKING WITH HVAC – A PRACTICAL LIST OF MUST-HAVE TOOLS AND THEIR USES 18
- Tools for Pressure Control and Refrigerant Handling 19
- Electrical Diagnostic Tools 19
- Tools for Working with Copper Pipes and Ducts 19
- Tools for Cleaning and Maintenance 20
- Why Having the Right Tools Matters 20
- Main Risks in HVAC Systems 21
- Safety Regulations in the U.S.: What You Need to Know 21
- Personal Protective Equipment (PPE): Never Work Without It 23
- Best Practices for Safe HVAC Work 23
- From Safety to Understanding HVAC Components 24

BOOK 2 ESSENTIAL COMPONENTS OF AN HVAC SYSTEM: A PRACTICAL GUIDE 25

2.1 THE COMPRESSOR: THE HEART OF THE SYSTEM – HOW IT WORKS, TYPES, AND COMMON ISSUES 26
- How an HVAC Compressor Works 26
- Types of HVAC Compressors 27
- Common HVAC Compressor Issues 27
- From Compression to Heat Exchange: The Role of the Condenser and Evaporator 30

2.2 THE CONDENSER AND THE EVAPORATOR: THE ROLE OF THE REFRIGERATION CYCLE – HOW THESE COMPONENTS TRANSFER HEAT 30
- How Heat Transfer Works in an HVAC System 30
- The Condenser: Where Heat is Released 31
- The Evaporator: Where Heat is Absorbed 31
- The Condenser and Evaporator: Working Together with Different Functions 32

2.3 THERMOSTATS AND ELECTRONIC CONTROLS – HOW THEY WORK AND HOW TO SET THEM CORRECTLY FOR OPTIMAL EFFICIENCY 34

How a Thermostat Works in an HVAC System 34
Types of Thermostats 34
How to Set a Thermostat for Optimal Efficiency 35
The Importance of Electronic Controls in HVAC Systems 35
From Temperature Control to Air Quality 36

2.4 AIR FILTERS AND INDOOR AIR QUALITY – FILTER MAINTENANCE AND THEIR IMPACT ON HEALTH AND SYSTEM PERFORMANCE 36
How Air Filters Work in an HVAC System 37
Types of Air Filters and Their Efficiency 37
Filter Maintenance: How Often Should You Replace Them? 38
The Impact of Air Quality on Health and Comfort 38

2.5 REFRIGERANTS AND THEIR ENVIRONMENTAL IMPACT – DIFFERENCES BETWEEN TYPES AND REGULATIONS ON REPLACEMENTS 39
The Role of Refrigerants in the HVAC Cycle 39
Types of Refrigerants and Their Environmental Impact 40
Regulations on Replacements and the Transition to New Refrigerants 40
The Impact of Refrigerant Choice on Efficiency and Operating Costs 42

BOOK 3 INSTALLING AN HVAC SYSTEM: FROM THEORY TO PRACTICE 44

3.1 CHOOSING THE RIGHT HVAC SYSTEM – NEEDS ANALYSIS, LOAD CALCULATION, AND SYSTEM SELECTION 45
Needs Analysis: What Type of System Do You Really Need? 45
Calculating Thermal Load: A Crucial Step 46
Types of HVAC Systems: Which One to Choose? 47
Energy Efficiency and Operating Costs 48
Installation Preparation: Essential Tools 48
Necessary Materials for Installation 49
Safety: Essential Precautions During Installation 50
Complete Installation Checklist 50

3.3 KEY STEPS FOR INSTALLING A RESIDENTIAL HVAC SYSTEM – PRACTICAL GUIDE WITH REAL EXAMPLES 51
Positioning and Mounting the Outdoor Unit 51
Installing the Indoor Unit and Connecting to Ductwork 52
Connecting the Refrigerant Lines and Leak Testing 52
Electrical Connections and Thermostat Setup 53
First Startup and Final Inspection 53

3.4 INSTALLING A COMMERCIAL HVAC SYSTEM: KEY DIFFERENCES AND CHALLENGES – STRATEGIES FOR MANAGING LARGE SPACES 54
Key Differences Between Residential and Commercial HVAC Systems 55
Key Phases of a Commercial HVAC Installation 55
Selecting the System and Planning the Layout 55
Installing the Outdoor and Indoor Units 56
Ductwork Connection and Airflow Balancing 56
Configuring the Control System and Automation 57

3.5 FINAL INSPECTION AND TESTING: ENSURING EVERYTHING WORKS PERFECTLY – POST-INSTALLATION CHECKLIST 57
Electrical Connection and Thermostat Check 58
Airflow Test and Duct Balancing 58
Refrigerant Pressure Verification and Leak Detection 59
Temperature Testing and Energy Efficiency Check 59
Final Post-Installation Checklist 60

BOOK 4 DIAGNOSING AND SOLVING COMMON HVAC SYSTEM PROBLEMS 61

4.1 THE HVAC SYSTEM WON'T TURN ON: CAUSES AND SOLUTIONS – ANALYSIS AND TESTING FOR DIAGNOSING THE ISSUE 62
Checking the Electrical Power Supply 62
Checking the Thermostat and Settings 63

Checking the Control Relay and Contactor 63
Checking the Fan Motor and Compressor 64
Checking Refrigerant Pressure 64

4.2 Weak or Insufficient Airflow – How to Check for Clogged Ducts, Fans, and Filters 65
Checking Air Filters: The Primary Cause of Weak Airflow 65
Checking the Fan Blower: Dirty Blades or Motor Issues 65
Inspecting Air Ducts: Blockages and Leaks 66
Evaporator Coil Freezing 67
Final Testing and Restoring Maximum Performance 67

4.3 The Compressor Won't Start: How to Test and Replace It If Necessary – Diagnostics and Repair 68
Checking Power Supply and Contactor 68
Testing the Start Capacitor and Start Relay 68
Checking the Compressor's Thermal Conditions 69
Testing the Compressor Windings for Continuity 69
Replacing the Compressor: When Is It Necessary? 70

4.4 Refrigerant Leaks: How to Detect and Repair Them Properly – Practical Techniques with Specialized Tools 71
Signs of a Refrigerant Leak 71
Techniques for Detecting Refrigerant Leaks 72
Repairing Refrigerant Leaks 72
Refrigerant Recharge and Final Testing 73
Impact of Refrigerant Leaks on the System and the Environment 73
Types of Noises and Their Causes 74
Identifying the Source of the Noise 75
Troubleshooting and Fixing Common Issues 76
When Noise Indicates a Serious Problem 76
Final Testing and Restoring Silent Operation 77

BOOK 5 PREVENTIVE MAINTENANCE AND ENERGY EFFICIENCY OPTIMIZATION 78

5.1 Cleaning and Replacing Air Filters: When and How to Do It Properly – Periodic Maintenance Checklist..... 79
When to Replace or Clean Air Filters? 79
How to Properly Replace an Air Filter 80
Benefits of Regular Filter Maintenance 80
Periodic Maintenance Checklist for HVAC Filters 81

5.2 Inspection and Maintenance of the Compressor and Fans – Techniques to Extend Their Lifespan 81
Compressor Inspection: Signs of Malfunction 82
Compressor Maintenance: Preventing Overheating 82
Condenser Fan Inspection and Cleaning 83
Evaporator Fan Balancing and Adjustment 84
Final Check and System Optimization 84

5.3 Thermostat Adjustment and Strategies to Reduce Energy Consumption – Optimizing Temperature Without Waste 85
Ideal Thermostat Settings for Each Season 85
Benefits of Programmable and Smart Thermostats 86
Avoiding Waste with Smart Temperature Adjustments 86
Balancing Temperatures Between Rooms 87
Monitoring Energy Consumption and Thermostat Maintenance 87

5.4 Improving Indoor Air Quality – Strategies and Solutions for Healthier Air 88
The Role of Air Filters in Indoor Air Quality 88
Humidity and Air Quality: The Role of Humidifiers and Dehumidifiers 89
Ventilation and Air Exchange: Why It's Essential 89
Air Purifiers: When Are They Necessary? 90
Controlling Indoor Pollution Sources 90

5.5 Planning Professional HVAC Maintenance: When to Call an Expert – How to Identify Critical Malfunctions 91
- Signs of Malfunctions That Require Professional Service 92
- Recommended Frequency for Professional Maintenance 92
- Advanced Diagnostic Tests Performed by Technicians 93
- When the Issue is Too Severe: Repair or Replace the System? 93
- How to Choose a Reliable HVAC Technician 94

BOOK 6 TOOLS AND TECHNIQUES FOR ADVANCED DIAGNOSTICS 95

6.1 How to Use a Multimeter to Test Electrical Components 96
- Structure and Function of a Multimeter 96
- Measuring Voltage in an HVAC System 97
- Testing Continuity and Resistance 97
- Testing a Start Capacitor 98
- Identifying HVAC Circuit Problems with a Multimeter 98

6.2 How to Measure Refrigerant Pressure with an HVAC Manifold Gauge 99
- Types of Pressures in HVAC Systems and Their Meaning 99
- Structure of an HVAC Manifold Gauge and How to Use It 100
- Reading Pressure and Diagnosing the System 100
- Advanced Testing: Pressure Differences and Superheat Measurement 101
- Common Mistakes in Pressure Measurement and How to Avoid Them 101

6.3 Analysis of Operating Parameters with a Thermal Camera 102
6.4 Duct Leakage Testing and How to Detect Air Leaks 105
6.5 Software and Apps for HVAC System Management and Monitoring 107

BOOK 7 COMMON REPAIRS: STEP-BY-STEP GUIDE 111

7.1 Replacing a Faulty Capacitor 112
- Symptoms of a Faulty Capacitor 112
- Testing the Capacitor with a Multimeter 112
- Removing the Faulty Capacitor 113
- Installing the New Capacitor 113
- System Test and Final Verification 114

7.2 Repairing a Stuck Fan 114
- Signs of a Stuck or Malfunctioning Fan 114
- Diagnosing the Cause of the Blockage 115
- Manually Unlocking and Cleaning the Fan 115
- Replacing the Fan Motor Bearings 116
- Testing the Fan Motor and Start Capacitor 116

7.3 Replacing a Faulty Relay 117
- Symptoms of a Faulty Relay 117
- Testing the Relay with a Multimeter 117
- Removing the Faulty Relay 118
- Installing the New Relay 118
- System Test and Final Verification 119

7.4 How to Repair an HVAC System with a Refrigerant Leak 119
- Signs of a Refrigerant Leak 120
- Detecting the Refrigerant Leak 120
- Repairing the Leak 121
- Recharging the Refrigerant and Testing the System 121
- Preventing Future Leaks 121

7.5 Repairing a Faulty Thermostat 122
- Signs of a Faulty Thermostat 122
- Diagnosing a Thermostat Malfunction 123

Repairing and Resetting the Thermostat 123
Replacing a Faulty Thermostat 124
Preventing Future Malfunctions 124

BOOK 8 WORKING IN THE HVAC INDUSTRY: CAREER AND CERTIFICATIONS 126

8.1 SKILLS REQUIRED TO BECOME AN HVAC TECHNICIAN 127
Understanding HVAC System Operations 127
Diagnostic and Problem-Solving Skills 127
Practical Skills in Installation and Maintenance 128
Knowledge of Electrical and Electronic Systems 128
Safety and Regulatory Compliance Skills 129

8.2 CERTIFICATIONS AND LICENSES REQUIRED TO WORK IN THE HVAC INDUSTRY 129
EPA Section 608 – Refrigerant Handling Certification 129
NATE – North American Technician Excellence 130
State-Specific Certifications and Local Licenses 131
OSHA – Workplace Safety Certification 131
Other Specialized Certifications 132

8.3 HOW TO START A CAREER OR BUSINESS IN HVAC 132
Training and Specialization 132
Obtaining Required Certifications 133
Working as an HVAC Technician: Opportunities and Career Paths 133
Starting an HVAC Business 134
Staying Updated and Growing in the Industry 135

8.4 MISTAKES TO AVOID AS AN HVAC TECHNICIAN 135
Ignoring a Complete Diagnosis of the Problem 135
Failing to Follow Safety Procedures 136
Incorrect Installation of Key Components 136
Underestimating the Importance of Preventive Maintenance 137
Poor Communication with Customers 137

8.5 GROWTH OPPORTUNITIES IN THE HVAC INDUSTRY AND ADVANCED SPECIALIZATIONS 138
Advanced Career Paths in the HVAC Industry 138
Specializing in Energy Efficiency and Low-Impact Systems 139
HVAC and Automation: Specializing in Smart Systems 139
Predictive Maintenance and Advanced Diagnostics 140
Career Opportunities in the Commercial and Industrial Sectors 140

BOOK 9 SMART HVAC: THE FUTURE OF CLIMATE CONTROL 141

9.1 INTRODUCTION TO SMART AND CONNECTED HVAC SYSTEMS 142
What Are Smart and Connected HVAC Systems? 142
Benefits of Smart HVAC Systems 142
Key Components of a Smart HVAC System 143
Integration with Smart Homes and Home Automation 144
The Future of Connected HVAC Systems 144

9.2 INTEGRATION WITH HOME AUTOMATION AND VOICE ASSISTANTS 145
How HVAC and Home Automation Integration Works 145
HVAC Control via Voice Assistants 146
Advanced Automations and Scheduling 146
Integration with Environmental Sensors and Smart Home Systems 147
The Future of HVAC and Home Automation Integration 147

9.3 REMOTE MONITORING AND PREDICTIVE MAINTENANCE 148
What Is Remote Monitoring and How Does It Work? 148
Predictive Maintenance: Preventing Failures Before They Happen 149
Benefits of Remote Monitoring for Technicians and Users 149

Tools and Technologies for Monitoring and Diagnostics 150
The Future of HVAC Monitoring: Automation and Artificial Intelligence 151
9.4 HVAC AND SUSTAINABILITY: SOLUTIONS TO REDUCE ENVIRONMENTAL IMPACT 151
The Importance of Sustainability in the HVAC Industry 152
Sustainable HVAC Technologies 152
Renewable Energy and HVAC 153
Regulations and Incentives for Green HVAC Systems 153
Strategies for More Sustainable HVAC Usage 154
9.5 HOW TO CHOOSE A MODERN AND ENERGY-EFFICIENT HVAC SYSTEM 154
Evaluating the Building's Heating and Cooling Needs 154
Types of Energy-Efficient HVAC Systems 155
The Importance of Energy Efficiency Ratings 155
Smart Technologies to Maximize Efficiency 156
Upfront Cost vs. Long-Term Savings 156

BOOK 10 CONCLUSION AND FINAL CHECKLIST 157
10.1 SUMMARY OF KEY CONCEPTS 158
How an HVAC System Works 158
The Importance of Maintenance and Diagnostics 159
Energy Efficiency and Choosing the Right System 159
Future Technologies and Smart HVAC 160
The Role of HVAC Technicians and the Importance of Training 160
10.2 ANNUAL HVAC MAINTENANCE CHECKLIST 161
Air Filter Inspection and Cleaning 161
Inspection and Cleaning of Condenser and Evaporator Coils 162
Refrigerant Level Check and Leak Detection 162
Thermostat and Control System Check 162
Inspection of Fans, Motors, and Drive Belts 163
10.3 ESSENTIAL TOOLS FOR EVERY HVAC TECHNICIAN 163
Hand Tools and Basic Equipment 164
Electrical Diagnostic Tools 164
Refrigerant Measurement Tools 165
Airflow and Indoor Air Quality Monitoring Tools 165
Specialized Tools and Advanced Technologies 166
10.4 RESOURCES AND RECOMMENDED READINGS 166
Essential HVAC Books for Technicians 167
Training Courses and Certifications 167
Websites and Online Resources 168
Apps and Software for HVAC Professionals 168
HVAC Industry Events and Conferences 169
10.5 NEXT STEPS: HOW TO CONTINUE ADVANCING IN THE HVAC INDUSTRY 169
Specializing in a Specific HVAC Field 170
Earning Advanced Certifications and Professional Qualifications 171
Using Technology to Improve Skills 171
Attending HVAC Events, Conferences, and Trade Shows 172
Building a Strong Career and Looking to the Future 172

CONCLUSION 174

BOOK 1
Introduction to HVAC: What You Need to Know Before Getting Started

1.1 What is an HVAC System and Why is It Essential?

If you've ever wondered how heating works in winter or air conditioning in summer, you've already encountered an HVAC system—perhaps without realizing it. **HVAC** stands for **Heating, Ventilation, and Air Conditioning**. This system is designed to regulate temperature, control humidity, and improve air quality inside buildings, ensuring a comfortable and safe environment for occupants.

But why is it so essential? Simple: without HVAC, indoor spaces would quickly become too hot, too cold, or unhealthy. An efficient HVAC system not only maintains a pleasant temperature but also filters air impurities, regulates humidity levels, and ensures proper air circulation. This is especially important not only in homes but also in offices, stores, restaurants, and hospitals, where air quality is crucial for people's well-being.

The Three Pillars of an HVAC System

An HVAC system is based on three primary functions that work together to create an optimal indoor climate:

1. Heating

Heating is essential for keeping indoor spaces warm during colder months. Heating systems can be powered by **gas, electricity, fuel oil**, or **renewable sources** like **geothermal heat pumps**. The main heating methods include:

- **Boilers and radiators** – Mainly used in older homes and commercial buildings.
- **Gas furnaces** – Common in the U.S., they heat air and distribute it through ductwork.
- **Heat pumps** – Efficient systems that can both heat and cool spaces using external energy.
- **Radiant floor heating** – A growingly popular solution that evenly distributes heat.

2. Ventilation

Ventilation is the process of removing stale air from a building and replacing it with fresh air. This is crucial for:

- **Eliminating excess humidity** and preventing mold growth.
- **Removing odors, dust, allergens, and pollutants.**
- **Maintaining an adequate oxygen level in enclosed spaces.**

Ventilation systems can be **natural** (like open windows) or **mechanical** (fans, air exchangers, exhaust ducts, and filters). In large commercial buildings, ventilation is often integrated into air treatment systems that filter and regulate airflow.

3. Air Conditioning

When temperatures rise, HVAC systems must cool the air to ensure comfort. Air conditioning works mainly through:

- **Centralized air conditioning units** – Used in large buildings to distribute cool air across multiple rooms.
- **Split air conditioners** – Common in homes and offices, consisting of an outdoor and an indoor unit.
- **Heat pumps** – In addition to heating, they can also cool spaces in summer.
- **Evaporative cooling** – A technique used in dry climates that lowers temperature by evaporating water.

Why Is an HVAC System Essential?

Now that we've covered the three pillars of HVAC, it's clear why a well-designed system is crucial. Here are the key reasons:

✓ **Thermal comfort** – Regulating temperature is vital for living and working in optimal conditions.

✓ **Health and safety** – Good air quality reduces allergies, respiratory issues, and prevents the buildup of CO_2 and other harmful gases.

✓ **Energy efficiency** – A well-maintained HVAC system lowers energy consumption and reduces costs.

✓ **Material and building preservation** – Controlling humidity and temperature prevents mold, corrosion, and structural deterioration.

✓ **Regulations and standards** – Many buildings must comply with strict air quality and energy efficiency regulations.

These aspects apply to both **residential and commercial** HVAC systems, although there are key differences between them. In the next chapter, we'll explore these differences and understand why choosing the right solution is essential based on the building's purpose.

1.2 Types of HVAC Systems: Residential vs. Commercial – Key Differences and Why They Matter

Now that we've covered what an HVAC system is and why it's essential, it's important to understand that **not all HVAC systems are the same**. Climate control for a home is vastly different from that of an office, restaurant, or shopping mall. **Choosing the right system based on the environment to be conditioned is crucial** for efficiency, comfort, and reliability.

In this chapter, we'll explore the main differences between **residential and commercial HVAC systems**, explaining why it's important to understand them—especially if you want to work in the HVAC industry or simply gain a better understanding of how your system functions.

Key Differences Between Residential and Commercial HVAC Systems

While both systems share the same goal—regulating temperature, humidity, and air quality—there are five fundamental differences between HVAC systems for homes and those for commercial buildings.

1. System Capacity and Power

◆ Residential HVAC – Designed for smaller spaces with lower thermal demand. A typical residential unit ranges from 1 to 5 tons of cooling capacity (BTU/h), which is sufficient to maintain comfort in home environments.

◆ Commercial HVAC – A commercial building has much higher demands, especially because of larger open spaces, a higher number of occupants, and additional heat sources (computers, machinery, lighting). Commercial HVAC units range from 5 to 50 tons or more, with centralized systems distributing air uniformly across multiple floors.

System Type	Capacity Range (BTU/h)	Typical Application	Cooling Capacity (Tons)
Residential HVAC	12,000 - 60,000	Homes, Apartments	1 - 5
Commercial HVAC	60,000 - 600,000+	Offices, Malls, Factories	5 - 50+

2. System Structure and Configuration

◆ Residential – A typical home system consists of a single outdoor unit (compressor and condenser) and an indoor unit (evaporator and fan). These units are often installed on the roof, in the backyard, or next to the house.

◆ Commercial – Commercial HVAC systems are typically modular, meaning they consist of multiple combined units, allowing scalability based on demand. These systems are often installed on the building's rooftop or in a separate mechanical room to save space and reduce noise pollution.

3. Ventilation and Air Distribution

◆ **Residential** – Air is distributed through smaller ductwork **with a** limited number of vents**, making airflow regulation simpler.**

◆ **Commercial** – Commercial HVAC systems handle a much larger volume of air, using advanced ventilation systems with larger ducts, multiple zones, and heat recovery systems. In large buildings, even air distribution across all rooms and floors is essential.

4. Maintenance Complexity and Repairs

◆ **Residential HVAC** – Easier to maintain, as components are easily accessible, and diagnostics can be performed with basic tools.

◆ **Commercial HVAC** – Much more complex, as it involves multiple units, a network of sensors, and centralized controls. Maintenance often requires specialized personnel and advanced diagnostic tools.

5. Energy Efficiency and Automation

◆ **Residential** – Home HVAC systems are generally **less sophisticated** in terms of energy efficiency. However, many modern homes now feature **smart thermostats** to optimize energy consumption.

◆ **Commercial** – Large buildings utilize **advanced energy management systems**, such as **BMS (Building Management Systems)**, which optimize HVAC performance based on **outdoor temperature, room occupancy, and other parameters**.

Why Is It Important to Know These Differences?

✓ **To make the right choice based on the building's needs** – An HVAC system that is too small or too large can lead to **energy waste or inefficiencies**.

✓ **To understand maintenance and necessary repairs** – A commercial HVAC system **requires more frequent and professional servicing**, while a residential unit **can often be maintained with some DIY knowledge**.

✓ **For those pursuing a career in HVAC** – Maintaining a **home system** is very different from servicing a **commercial installation**. Understanding these differences is **essential for any professional HVAC technician**.

✓ **To improve energy efficiency and reduce costs** – Knowing how your HVAC system works helps **optimize energy consumption**, reducing **waste and utility bills**.

1.3 How Does an HVAC System Work? – The Thermodynamic Cycle Explained Simply

Now that we've covered the differences between residential and commercial HVAC systems, it's time to dive deeper into how an HVAC system actually works. While it may seem like a complex and technical topic, the reality is that all HVAC systems follow the same basic principles.

Imagine a scorching summer day—you walk into your house and turn on the air conditioner. Within minutes, the room is cooler and more comfortable. But have you ever wondered what happens behind the scenes? Or, on a freezing winter day, you press the heating button, and soon your home feels warm and cozy. How does this heat transfer take place?

The answer lies in the thermodynamic cycle—a process that moves heat from one place to another to maintain the desired temperature. Let's break it down into four essential phases.

The 4 Phases of the Thermodynamic Cycle

An HVAC system operates on a simple principle: heat always moves from a warmer area to a cooler one. To facilitate this process, the system uses a special fluid called refrigerant, which changes from liquid to gas and back again, absorbing and releasing heat along the way.

Compression: The Refrigerant is Compressed and Heats Up

Everything starts with the compressor, the heart of an HVAC system. This component receives the refrigerant in cold gas form and compresses it at high pressure. When a gas is compressed, its temperature increases—just like when you pump air into a bicycle tire, and the valve gets hot.

Condensation: Heat is Released into the Outdoor Environment

The hot, high-pressure gas then moves into the condenser, located in the outdoor unit of the HVAC system. Here, the refrigerant releases its heat to the surrounding air, cooling down and returning to a liquid state. It's the same principle as a hot cup of coffee—left in the open air, the heat disperses, and the coffee cools down.

Expansion: The Refrigerant Cools Down Rapidly

Once the refrigerant is in liquid form, it passes through the expansion valve, a small device that controls the refrigerant flow and reduces its pressure. When pressure decreases, the liquid cools drastically. It's the same effect as spraying deodorant—the gas inside expands rapidly and feels cold when it touches the skin.

Evaporation: The Indoor Air is Cooled

At this stage, the cold refrigerant enters the evaporator, located inside the building. Here, the refrigerant absorbs heat from the indoor air, cooling the room and transforming back into a gas. The fan in the indoor unit then distributes the cooled air throughout the space. The gas refrigerant then cycles back to the compressor, and the process repeats.

And What About Heating? The Cycle Works in Reverse!

If air conditioning removes heat from the indoor space and expels it outside, heating works in the exact opposite way.

Heat pumps, for example, reverse the refrigerant cycle, extracting heat from the outdoor air and bringing it inside. Even in winter, outdoor air contains some heat, which can be captured and used to warm indoor spaces.

This makes heat pumps much more energy-efficient than traditional combustion-based heating systems since they transfer heat rather than generate it directly.

The Role of Diagnostic Tools in HVAC Maintenance

Understanding the thermodynamic cycle is crucial not just for installing or repairing an HVAC system but also for accurate diagnostics.

For example, an **HVAC technician** must be able to:

- ✓ **Measure refrigerant pressure**
- ✓ **Check temperatures at various cycle points**
- ✓ **Evaluate compressor efficiency**
- ✓ **Detect refrigerant leaks**

To do this, they use specialized tools such as:

- ◆ **HVAC manifold gauges** – To measure refrigerant pressure
- ◆ **Infrared thermometers** – To check temperature levels
- ◆ **Leak detectors** – To identify refrigerant leaks

Mastering these tools is essential for ensuring the system operates **efficiently and safely**.

1.4 Essential Tools and Equipment for Working with HVAC – A Practical List of Must-Have Tools and Their Uses

After understanding how an HVAC system works, it's time to discuss the **tools needed to work efficiently**. Without the right equipment, even the most experienced technician can face difficulties. For beginners, knowing the **fundamental tools** is the first step to performing **installations, maintenance, and repairs** with precision and safety.

If you've ever tried fixing something at home without the right tools, you know how frustrating it can be. The same applies to HVAC systems: **each component requires specific tools** for testing, adjusting, or replacing. Some of these tools are **common** and found in any toolbox, while others are **specialized** and essential for daily HVAC work.

Basic Tools: The Essential Kit

Every HVAC technician should have a set of basic hand tools to handle the most common tasks.

A multi-bit screwdriver is essential for opening system panels and working on electrical connections. Adjustable pliers and wrenches are used for tightening fittings and pipes. A cordless drill allows for fast unit installation and bracket mounting.

A measuring tape is crucial for measuring pipes and installation spaces. A tubing cutter ensures precise cutting of copper pipes, while an Allen wrench set is needed for adjusting system valves. A LED flashlight is indispensable for working in dark spaces, such as attics and mechanical rooms.

Tools for Pressure Control and Refrigerant Handling

HVAC systems rely on proper refrigerant management and pressure control. The following tools are essential:

A manifold gauge set is crucial for measuring refrigerant pressure in the system. Without it, verifying system performance or detecting potential leaks would be impossible.

Refrigerant scales are used to precisely measure the amount of refrigerant added or recovered. Unlike a simple refill, ensuring the exact refrigerant charge is essential to avoid system malfunctions.

A leak detector is another key tool. Even a small leak can compromise system efficiency, and in the U.S., laws require immediate repair of leaks in commercial HVAC systems.

Electrical Diagnostic Tools

An HVAC system isn't just about pipes and refrigerant—it also includes an electrical circuit that controls fans, compressors, and sensors.

A digital multimeter is essential for checking voltage, continuity, and resistance in electrical components. If an HVAC system won't turn on, the issue is often electrical, such as a blown fuse or a faulty relay.

To test capacitors, which help motors start, a capacitance meter is used. Faulty capacitors are one of the most common causes of HVAC system failures, so having this tool helps diagnose issues quickly.

An infrared thermometer is useful for checking the temperature of pipes and components without making direct contact. This helps identify temperature fluctuations, which may indicate blockages or refrigerant issues.

Tools for Working with Copper Pipes and Ducts

Copper pipes are a core element of HVAC systems. To work with them, these tools are essential:

A flaring tool is used to create airtight connections between copper pipes. A tube bender allows pipes to be shaped without damage.

For ductwork, sheet metal snips are useful for cutting and shaping ventilation ducts. In commercial systems, ventilation often involves large metal duct sections, and precise cutting ensures optimal airflow.

Tools for Cleaning and Maintenance

HVAC systems work best when they are clean. Dust, dirt, and debris reduce efficiency and cause failures. Proper maintenance requires:

A compressed air blower helps clear dust from electrical components and condenser coils. A wet/dry vacuum is useful for removing debris and liquids from indoor and outdoor units.

Evaporator and condenser coils require regular cleaning. For this, coil brushes and chemical cleaners are used to remove dirt and buildup that can reduce system efficiency.

Why Having the Right Tools Matters

Having the right tools is not just about efficiency, but also about safety. Using the wrong equipment can lead to malfunctions, component damage, or even accidents. Many HVAC tools handle high voltage, high pressure, and hazardous chemicals, making proper usage crucial.

1.5 Safety First: Essential Precautions Before Working on an HVAC System – Regulations, PPE, and Best Practices to Avoid Accidents

Working on an HVAC system is not like fixing a regular household appliance. HVAC systems operate with high voltage, pressurized refrigerant gases, and moving mechanical components, making them potentially dangerous if not handled carefully. Before working on any system, it is essential to understand and follow safety regulations, use the correct personal protective equipment (PPE), and apply best practices to prevent accidents.

In the United States, anyone working in the HVAC industry must also be familiar with OSHA (Occupational Safety and Health Administration) regulations and comply with EPA (Environmental Protection Agency) guidelines for the safe handling of refrigerants. Failure to follow these precautions can lead to accidents, system failures, and legal penalties.

Main Risks in HVAC Systems

An HVAC system may appear harmless from the outside, but it contains electrical, chemical, and mechanical hazards. Electric shocks are among the most common risks, especially when working on compressors, capacitors, and fans. Even low voltage can be dangerous if the right precautions are not taken.

Refrigerants used in HVAC systems are pressurized chemicals that, if inhaled or improperly released, can cause respiratory issues and environmental damage.

Moving parts, such as fans and compressors, pose a serious risk to hands and fingers if the system is not properly turned off before servicing. Additionally, contact with overheated surfaces, such as coils and piping, can cause burns if the proper protective gear is not used.

Safety Regulations in the U.S.: What You Need to Know

In the United States, workplace safety is regulated by two main agencies:

- **OSHA** (Occupational Safety and Health Administration) – Establishes workplace safety standards, including electrical hazard prevention and PPE usage requirements.
- **EPA** (Environmental Protection Agency) – Enforces strict regulations on handling and recovering refrigerants to prevent harmful emissions.

Before working with refrigerants, technicians must obtain an EPA certification (Section 608 Clean Air Act Certification). Without this certification, it is illegal to purchase or handle refrigerants, and violations can result in severe fines. Additionally, refrigerant transport and disposal must follow strict guidelines to minimize environmental impact.

How To Obtain EPA Section 608 Certification

	Step	Description
1	Step 1: Understand the Certification Types	There are four types of EPA Section 608 certification: Type I (Small Appliances), Type II (High-Pressure Systems), Type III (Low-Pressure Systems), and Universal Certification (all types). Choose the one that fits your needs.
2	Step 2: Choose an Approved Testing Organization	Find an EPA-approved testing provider. Many community colleges, trade schools, and HVAC organizations offer certification exams.
3	Step 3: Study for the Exam	Study the official EPA guidelines, refrigerant handling procedures, and safety regulations. Many online resources and practice exams are available.
4	Step 4: Schedule and Take the Exam	Register for the test and take the exam at an approved location or online, if available. The test covers environmental regulations, safety procedures, and refrigerant handling.
5	Step 5: Receive Your Certification	If you pass the exam, you will receive your official EPA Section 608 certification, which allows you to purchase and handle refrigerants legally in the U.S.

Personal Protective Equipment (PPE): Never Work Without It

An HVAC technician must always wear the appropriate PPE to ensure safety.

- **Insulated gloves** – Protect against electrical shocks.
- **Chemical-resistant gloves** – Prevent direct contact with refrigerants.
- **Safety goggles** – Shield the eyes from sparks, debris, and chemical splashes.
- **Respirator mask with chemical vapor filters** – Prevents inhalation of **toxic gases** while working with refrigerants.
- **Slip-resistant and anti-static footwear** – Reduces the risk of electric shocks and workplace injuries.

An experienced HVAC technician never relies on luck—wearing proper PPE can be the difference between a safe job and a preventable accident.

Best Practices for Safe HVAC Work

- **Always disconnect power before servicing a system.** Even if a unit appears off, never assume it is completely de-energized. Use a voltage tester to confirm that no power is flowing before touching electrical components.
- **Proper refrigerant recovery is mandatory.** Releasing refrigerant into the atmosphere is not only dangerous but also illegal under EPA regulations. **HVAC recovery machines** are used to safely capture, store, or dispose of refrigerants according to legal requirements.
- **Ensure proper ventilation.** Some refrigerants, if inhaled in enclosed spaces, can cause asphyxiation. Always work in well-ventilated areas or use exhaust fans to prevent respiratory issues.
- **Use proper lifting techniques for heavy HVAC equipment.** Compressors and outdoor units should never be lifted improperly. Many HVAC injuries occur due to incorrect lifting methods. Using lifting carts, hoists, or team-lifting helps prevent back and limb injuries.
- **Perform a final safety check before starting a new system.** Ensure that:
  - **All electrical connections are secure.**

- There are no refrigerant leaks.
- The wiring is in perfect condition.

    This significantly reduces the risk of **system failures and workplace accidents**.

From Safety to Understanding HVAC Components

Now that we have established the fundamental safety rules for working with HVAC systems, it's time to explore the key components of an HVAC system.

The first critical component is the compressor, often referred to as the heart of the system, as it is responsible for compressing refrigerant and driving the entire cooling cycle. Without a properly functioning compressor, an HVAC system cannot operate.

In the next chapter, we'll examine how a compressor works, the different types available, and the most common issues that can cause it to fail.

BOOK 2
Essential Components of an HVAC System: A Practical Guide

2.1 The Compressor: The Heart of the System – How It Works, Types, and Common Issues

Every HVAC system depends on its beating heart—the compressor, the component responsible for circulating refrigerant throughout the cooling and heating cycle. Without a functioning compressor, an HVAC system cannot effectively transfer heat, making it impossible to cool or heat indoor spaces.

To understand how an HVAC system operates, it's essential to grasp the compressor's role in the refrigeration cycle—the process through which heat is absorbed from one space and released elsewhere. In the previous chapter, we discussed the importance of safety when working with HVAC components. Now, let's dive into how a compressor works, the main types available, and common issues that can affect its performance.

How an HVAC Compressor Works

The compressor's function is to increase the refrigerant's pressure and circulate it through the system. In the refrigeration cycle, the refrigerant enters the compressor as a cold, low-pressure gas. The compressor compresses and heats it, transforming it into a high-pressure, high-temperature gas. This step is crucial because it allows the refrigerant to release the absorbed heat when it reaches the condenser.

To visualize this, imagine pumping air into a bicycle tire: as you increase the air pressure inside the tube, it heats up. The compressor works in a similar way, pressurizing the refrigerant to make it more energy-efficient for heat transfer.

Once the refrigerant is compressed, it is pushed toward the condenser, where the excess heat is released into the external environment. From there, the refrigerant cools down, turns back into a liquid, and continues through the system, ready to repeat the cycle.

Types of HVAC Compressors

There are several types of compressors used in HVAC systems, each with its own characteristics and applications. The choice of the right compressor depends on system size, efficiency requirements, and operating conditions.

- Reciprocating Compressors (Piston-Type) – Among the most common in residential and small commercial HVAC systems, these use a piston to compress the refrigerant inside a cylinder. They are reliable and easy to repair but less efficient than other models.
- Scroll Compressors – A popular choice for residential and commercial systems, these use two spiral-shaped scrolls to compress the refrigerant without pistons, ensuring quieter operation with fewer vibrations. They are more efficient than reciprocating models and require less maintenance.
- Rotary Compressors – These use rotating blades to compress the refrigerant. They are compact, efficient, and ideal for medium-sized residential and commercial applications.
- Screw Compressors (Screw-Type) – Primarily used in commercial and industrial systems, these operate with two helical screws that continuously compress the refrigerant. They are more efficient than reciprocating compressors and suitable for high-load applications.
- Centrifugal Compressors – Common in large industrial systems, these utilize centrifugal force to compress the refrigerant. They are highly efficient for large-scale operations but require specialized maintenance.

Common HVAC Compressor Issues

The compressor is one of the most stressed components of an HVAC system, making it prone to various failures. When a compressor malfunctions, the entire HVAC system suffers, leading to inefficiencies, increased energy consumption, and system failures.

- Overheating – If the compressor runs too long without adequate cooling, it can become damaged. This may be caused by poor ventilation, low refrigerant levels, or clogged air filters that force the system to work harder.

- Refrigerant Leaks – If the system has a leak in the piping, the compressor must work harder to maintain proper pressure, increasing the risk of failure. Detecting and repairing leaks is critical for extending the compressor's lifespan.
- Electrical Issues – Faulty wiring, a defective capacitor, or a damaged relay can prevent the compressor from starting properly. Regular electrical inspections and replacing defective components help maintain performance.
- Blockages Due to Contaminants – Impurities in the refrigerant, such as metal particles or dirt, can damage internal parts of the compressor. Keeping filters and lubricants clean prevents these failures.
- Unusual Noises – If the compressor makes strange sounds, such as buzzing, clicking, or metallic knocking, it could indicate a mechanical issue or an internal motor problem.

How To Diagnose A Non-Starting HVAC Compressor

	Quick Test	Solution
1	Check Power Supply	Ensure the system is receiving power. Check the circuit breaker and fuses.
2	Inspect the Capacitor	Use a multimeter to test if the capacitor is working. Replace if faulty.
3	Test the Start Relay	Test the start relay with a multimeter. A faulty relay will prevent the compressor from turning on.
4	Look for Refrigerant Leaks	Check for signs of refrigerant leakage, such as oil stains or hissing sounds. Repair leaks and recharge refrigerant.
5	Measure Compressor Resistance	Use an ohmmeter to measure resistance between compressor terminals. Unusual readings may indicate a burnt-out motor.
6	Listen for Unusual Noises	If the compressor makes loud buzzing, clicking, or knocking sounds, it may have internal damage. Professional repair or replacement may be needed.

From Compression to Heat Exchange: The Role of the Condenser and Evaporator

Now that we understand the importance of the compressor and the issues that can compromise its efficiency, it's time to explore two other critical components of the HVAC system: the condenser and evaporator.

These elements work together to transfer heat between the indoor and outdoor environments, completing the refrigeration cycle.

In the next chapter, we'll examine how the condenser and evaporator function, their role in cooling and heating air, and the most common problems affecting their performance.

2.2 The Condenser and the Evaporator: The Role of the Refrigeration Cycle – How These Components Transfer Heat

In the previous chapter, we explored how the compressor serves as the heart of an HVAC system, responsible for compressing the refrigerant and moving it through the circuit. However, without the condenser and evaporator, heat transfer would not be possible. These two components work together to extract heat from one environment and release it into another, completing the refrigeration cycle.

Understanding how the condenser and evaporator function is not only useful for grasping the HVAC system's operation, but it is also essential for diagnostics and maintenance. A problem with either of these components can severely affect the system's ability to cool or heat a space.

How Heat Transfer Works in an HVAC System

An HVAC system does not generate cold air directly; instead, it works by transferring heat from one place to another. The heat present in an indoor space is absorbed, transported via the refrigerant, and released into an external or internal area, depending on whether the system is in cooling or heating mode.

The condenser and evaporator are the two components responsible for this process. The evaporator absorbs heat from indoor air, while the condenser releases it into the external environment. This heat exchange occurs due to the phase change of the refrigerant, which alternates between liquid and gas states, allowing the system to function efficiently.

The Condenser: Where Heat is Released

The condenser is located in the outdoor unit of an HVAC system, typically placed outside for residential units or on the roof for commercial systems. After the refrigerant is compressed and heated, it enters the condenser as a high-pressure, high-temperature gas.

Inside the condenser, the refrigerant passes through metal coils, which help dissipate heat. A fan forces outdoor air over these coils, allowing the refrigerant to release its heat into the external environment. As the refrigerant cools down, it returns to a liquid state, ready for the next stage of the cycle.

If the condenser does not function properly, the system fails to release heat efficiently, leading to compressor overheating, increased energy consumption, and reduced system performance. A dirty or clogged condenser can restrict airflow, preventing proper heat dissipation.

The Evaporator: Where Heat is Absorbed

The evaporator is located inside the building and is responsible for absorbing heat from indoor air. Once the liquid refrigerant exits the condenser, it passes through the expansion valve, which reduces its pressure and temperature drastically.

As the cold refrigerant enters the evaporator coils, it absorbs heat from the surrounding air. This process causes the refrigerant to evaporate, turning into a low-pressure, low-temperature gas. The cooled air is then distributed throughout the space via the ventilation system, lowering the indoor temperature.

If the evaporator becomes dirty or clogged, the system fails to absorb heat properly, resulting in reduced efficiency. Additionally, when air humidity is high, condensation can accumulate on the evaporator coils, requiring proper drainage to prevent mold or blockages in the system.

The Condenser and Evaporator: Working Together with Different Functions

Although the condenser and evaporator have opposite roles, they must function in perfect balance to ensure an efficient refrigeration cycle. A failure in either component immediately affects the other.

- If the condenser cannot release heat efficiently, the refrigerant will not cool down properly before reaching the evaporator.
- If the evaporator is blocked, the refrigerant will not absorb heat effectively, reducing the system's cooling capacity.

One of the most common causes of imbalance between the condenser and evaporator is an incorrect refrigerant charge.

- If the refrigerant level is too low, the compressor will work harder to maintain pressure, increasing the risk of failure.
- If the refrigerant level is too high, it can cause flow problems and reduced energy efficiency.

Another frequent issue is the formation of ice on the evaporator, which can block airflow and drastically reduce performance. This often happens when filters are dirty or airflow is restricted.

How To Identify And Fix A Frozen Evaporator Coil

	Warning Signs	Solution
1	Reduced Airflow	Check and replace clogged air filters to improve airflow.
2	Ice Buildup on the Coils	Turn off the system and let the ice melt. Do not chip away the ice manually.
3	Warm Air from Vents	Ensure refrigerant levels are adequate. Low refrigerant can cause freezing.
4	Unusual Hissing or Bubbling Sounds	Inspect and clean the evaporator coil to remove dirt and dust buildup.
5	High Energy Bills	Verify proper thermostat settings and avoid running the AC at excessively low temperatures.

2.3 Thermostats and Electronic Controls – How They Work and How to Set Them Correctly for Optimal Efficiency

In the previous chapter, we explored how the condenser and evaporator work together to facilitate heat transfer and maintain a comfortable indoor environment. However, even the best HVAC system can become inefficient without proper control mechanisms. Thermostats and electronic controls play a crucial role in managing temperature, optimizing energy consumption, and ensuring overall comfort in both residential and commercial buildings.

In the United States, where energy costs continue to rise and efficiency is a growing priority, knowing how to properly set and use a thermostat can mean the difference between a manageable utility bill and excessive energy consumption.

How a Thermostat Works in an HVAC System

The thermostat serves as the main control point for an HVAC system. Its function is to measure the indoor temperature and send signals to the system to activate or deactivate heating or cooling as needed.

Thermostats operate based on an internal temperature sensor. When the detected temperature deviates from the user-set value, the thermostat sends a signal to the HVAC system, instructing it to start the compressor and begin the heating or cooling cycle. Once the desired temperature is reached, the thermostat shuts off the system to prevent unnecessary energy use.

There are several types of thermostats, each with unique features that impact efficiency and climate control management.

Types of Thermostats

- Electromechanical Thermostats – These are the simplest and most traditional models. They function using a bimetallic strip that expands and contracts with temperature changes, opening or closing the electrical circuit of the HVAC system. While they are affordable and reliable, their accuracy is limited, leading to larger temperature fluctuations.

- Programmable Thermostats – These allow users to set different temperatures for various times of the day. This means the HVAC system can automatically reduce heating or cooling when the house is empty and restore the desired temperature before occupants return. This improves efficiency and reduces energy consumption.
- Smart Thermostats – The most advanced solution, these are Wi-Fi-enabled and can be controlled via smartphone, tablet, or voice assistants like Alexa and Google Assistant. Using artificial intelligence and occupancy sensors, smart thermostats can learn user habits and automatically adjust temperatures to maximize comfort and energy savings.

How to Set a Thermostat for Optimal Efficiency

Properly configuring a thermostat is essential to ensuring efficient HVAC operation without wasting energy.

In the United States, the Department of Energy recommends setting thermostats to 78°F (25°C) in summer when at home and 85°F (29°C) when away. In winter, the ideal settings are 68°F (20°C) when at home and 60°F (16°C) when sleeping or away.

A common mistake is lowering the thermostat drastically, expecting the house to cool down faster. However, HVAC systems operate at a fixed rate, so setting a much lower temperature does not speed up cooling—it only wastes energy.

Another important factor is thermostat placement. If installed in a sunlit area or near a heat source (such as a kitchen), temperature readings can be inaccurate, causing inefficient heating and cooling cycles. The thermostat should be placed in a central part of the home, away from windows and exterior doors.

The Importance of Electronic Controls in HVAC Systems

In addition to thermostats, modern HVAC systems utilize a range of advanced electronic controls to optimize performance and improve efficiency. These include:

- Humidity sensors – Adjust system operation based on indoor humidity levels.
- Air quality sensors – Monitor and improve indoor air conditions.
- Remote monitoring systems – Allow users to control and adjust HVAC settings from anywhere.

Many commercial buildings use BMS (Building Management Systems), which integrate HVAC controls with lighting and other systems to enhance the building's overall energy efficiency.

In residential systems, advanced controls may include variable-speed fans, which automatically adjust airflow to reduce energy consumption without sacrificing comfort.

From Temperature Control to Air Quality

Setting up the thermostat and electronic controls correctly is only one part of achieving an efficient HVAC system. Another critical factor is indoor air quality, which primarily depends on air filters.

A dirty or clogged filter can severely reduce HVAC efficiency, leading to higher energy costs and poorer indoor air quality.

For this reason, in the next chapter, we will explore how to maintain clean air filters and why proper filter maintenance is essential for both system performance and respiratory health.

2.4 Air Filters and Indoor Air Quality – Filter Maintenance and Their Impact on Health and System Performance

After discussing how thermostats and electronic controls regulate temperature and improve HVAC efficiency, it's time to focus on a crucial but often overlooked component—air filters.

An HVAC system may be perfectly calibrated, with a well-functioning condenser and evaporator, but if the air filters are not properly maintained, the entire system suffers.

Air filters serve a dual essential function:

1. They protect the HVAC system from dust and debris.
2. They improve indoor air quality by capturing allergens, particulates, and pollutants.

Neglecting air filter maintenance reduces system performance, increases energy consumption, and raises the risk of malfunctions. More importantly, it can negatively affect the health of those living or working in the building.

How Air Filters Work in an HVAC System

Air filters are positioned in the return air duct of the HVAC system, before the air is treated and recirculated indoors. Their role is to trap airborne particles, preventing them from entering the ducts, evaporator, and other system components.

As air is drawn through the ducts, it first passes through the filter, where dust, pollen, pet dander, mold spores, and other contaminants are captured. A clean filter allows air to flow freely, ensuring proper circulation and reducing strain on the fan motor.

A dirty or clogged filter, however, restricts airflow, forcing the HVAC system to work harder and consume more energy.

Types of Air Filters and Their Efficiency

Not all HVAC filters are the same. In the United States, their efficiency is classified by the MERV (Minimum Efficiency Reporting Value) scale, which ranges from 1 to 20. A higher MERV rating indicates a greater filtration capability.

- Fiberglass Filters (MERV 1-4) – The most affordable and common in residential systems. They block large particles like dust and pollen but are less effective against smaller allergens.
- Pleated Filters (MERV 5-13) – Provide better protection, capturing mold spores and bacteria. These are recommended for allergy sufferers and those with respiratory conditions.

- HEPA Filters (MERV 17-20) – The most advanced filters, used in hospitals, laboratories, and sensitive environments. They remove up to 99.97% of airborne particles, including viruses and ultra-fine pollutants.

Filter Maintenance: How Often Should You Replace Them?

To maintain HVAC efficiency, air filters should be cleaned or replaced regularly. The replacement frequency depends on factors like filter type, system usage, and indoor air quality.

- In residential HVAC systems, a standard filter should be replaced every 1-3 months.
- Homes with pets or allergy sufferers may require more frequent changes.
- In commercial HVAC systems, filter maintenance is even more critical, often requiring monthly checks and replacements, especially in environments with high dust or pollution levels.

A dirty filter not only reduces air quality, but it can also increase HVAC energy consumption by up to 15%, as the fan must work harder to push air through the clogged filter.

The Impact of Air Quality on Health and Comfort

Indoor air quality plays a crucial role in overall health and well-being. A well-maintained HVAC system with clean filters helps to reduce allergens, fine dust, and airborne pathogens, improving breathing conditions and preventing health issues such as:

✓ Allergies

✓ Asthma

✓ Respiratory infections

In the United States, where people spend up to 90% of their time indoors, air quality is closely linked to public health. The CDC (Centers for Disease Control and Prevention) and EPA (Environmental Protection Agency) recommend regular HVAC maintenance to minimize exposure to indoor pollutants.

Clean air also improves workplace productivity and comfort. Studies show that better air quality can boost efficiency and reduce sick days.

2.5 Refrigerants and Their Environmental Impact – Differences Between Types and Regulations on Replacements

After exploring the importance of air quality and filter maintenance for HVAC efficiency, it's crucial to examine another key element: refrigerants.

Refrigerants are chemical substances that absorb and transfer heat within the heating and cooling cycle of an HVAC system. Over time, their composition has drastically changed due to environmental regulations and the need to reduce climate impact. In the United States, the Environmental Protection Agency (EPA) has banned or regulated the use of certain refrigerants to minimize ozone depletion and limit greenhouse gas emissions.

Choosing the right refrigerant is not just about system compatibility—it also has direct implications on environmental sustainability, operational costs, and compliance with federal regulations.

The Role of Refrigerants in the HVAC Cycle

The refrigerant is the working fluid that enables heat transfer between the condenser and evaporator. As the compressor pressurizes the refrigerant, its temperature rises, allowing it to release heat into the outdoor environment through the condenser. Once cooled and depressurized, the refrigerant moves into the evaporator, where it absorbs heat from indoor air and transforms back into gas.

Without a proper refrigerant, an HVAC system cannot function efficiently. Low refrigerant levels or the use of an incompatible fluid can lead to reduced efficiency, higher energy consumption, and potential compressor failure.

Types of Refrigerants and Their Environmental Impact

Over the years, the HVAC industry has used various types of refrigerants, some of which were later banned due to their harmful environmental effects.

- CFCs (Chlorofluorocarbons) – Examples include R-12, one of the earliest refrigerants used in HVAC systems. Banned under the 1987 Montreal Protocol due to their high ozone depletion potential (ODP).
- HCFCs (Hydrochlorofluorocarbons) – R-22 was widely used as a replacement for CFCs but still posed an ozone threat. The EPA banned R-22 production and imports in 2020, although older systems still use it and may require retrofits or replacements.
- HFCs (Hydrofluorocarbons) – Examples include R-410A and R-134a, which became the standard for modern HVAC systems. They contain no chlorine, meaning zero ozone depletion, but have a high global warming potential (GWP). Due to this, HFCs are now being phased out in favor of greener alternatives.
- HFOs (Hydrofluoroolefins) – Examples include R-32 and R-454B, representing the future of HVAC refrigerants. They are designed for low GWP and improved energy efficiency, making them eco-friendly alternatives.

Refrigerant Comparison Table

Refrigerant	Type	Ozone Depletion Potential (ODP)	Global Warming Potential (GWP)	Regulatory Status
R-22	HCFC	0.05	1810	Banned in the US (since 2020)
R-410A	HFC	0.0	2088	Being phased out due to high GWP
R-32	HFO	0.0	675	Low-GWP alternative, gaining popularity
R-454B	HFO	0.0	466	Next-gen refrigerant, low GWP, EPA-approved

Regulations on Replacements and the Transition to New Refrigerants

In the United States, refrigerant regulations are strictly monitored by the EPA and the Clean Air Act, which restrict the use of high-impact substances.

With the phase-out of R-22, many HVAC owners face the decision of whether to convert existing systems to alternative refrigerants or replace them entirely with newer, compliant models.

Starting in 2025, the EPA will impose further restrictions on HFCs with high GWP, encouraging the transition to R-32 and R-454B, which offer higher energy efficiency and lower environmental impact.

Regulations also include strict rules on refrigerant recovery and recycling, prohibiting atmospheric release and requiring specialized equipment for safe refrigerant recovery and reuse. HVAC technicians must be EPA Section 608 certified to handle and replace refrigerants legally, ensuring compliance with environmental laws.

What The R-22 Phase-Out Means And System Compatibility

	Key Aspect	Explanation
1	Why is R-22 Phased Out?	R-22 is being phased out due to its ozone depletion potential (ODP) and high global warming potential (GWP). The EPA has banned its production and import in the U.S. since 2020.
2	How to Identify If Your System Uses R-22	Check your system's nameplate (usually on the outdoor unit) for the refrigerant type. If it says 'R-22,' your system uses the banned refrigerant.
3	Replacement Options	You can retrofit your system to use alternative refrigerants (such as R-407C or R-422D), but efficiency and performance may be affected.
4	Converting to a New Refrigerant	Some systems can be modified to use R-410A or newer low-GWP refrigerants. This requires component upgrades and professional servicing.
5	When to Replace Your HVAC System	If your HVAC system is over 15 years old or requires frequent repairs, upgrading to a modern, energy-efficient system with an approved refrigerant (R-32, R-454B) is recommended.

The Impact of Refrigerant Choice on Efficiency and Operating Costs

Beyond environmental considerations, the choice of refrigerant directly impacts HVAC operating costs and system performance.

Modern refrigerants like R-32 not only reduce emissions but also require less volume to achieve the same cooling performance, improving energy efficiency and reducing operating costs.

In commercial HVAC systems, the choice of refrigerant also affects maintenance requirements and system longevity. Some refrigerants have chemical properties that reduce wear and tear on internal components, extending the lifespan of compressors and condensers.

For those installing a new HVAC system or replacing an existing unit, it's crucial to consider not just the upfront cost but also long-term expenses related to energy consumption and refrigerant availability.

BOOK 3
Installing an HVAC System: From Theory to Practice

3.1 Choosing the Right HVAC System – Needs Analysis, Load Calculation, and System Selection

After understanding the role of refrigerants and their environmental impact, it's time to address one of the most important decisions for anyone installing or replacing an HVAC system: choosing the right system.

A properly sized HVAC system ensures comfort, energy efficiency, and longevity.

- A unit that is too large will consume more energy than necessary, with frequent on/off cycles that can shorten its lifespan.
- A unit that is too small will work excessively without ever reaching the desired temperature, increasing operational costs and reducing indoor comfort.

To avoid these issues, the first step is to determine the building's thermal load, an essential factor in selecting a system that meets the actual climate control needs.

Needs Analysis: What Type of System Do You Really Need?

The choice of an HVAC system depends on several factors, including:

✓ Building size

✓ Local climate

✓ Thermal insulation

✓ Number of occupants

In the United States, an HVAC system in a cold-winter region like the Midwest will have very different energy requirements compared to a home in hot regions like Florida.

- A well-insulated building will require less energy to maintain temperature.
- A home with old windows and poor insulation will need a more powerful system.

Another key consideration is building usage.

- A residential home has different needs compared to an office or retail space.
- Commercial spaces require adequate ventilation and an HVAC system capable of handling heat from electronic equipment and large crowds.

Calculating Thermal Load: A Crucial Step

The thermal load represents the amount of energy needed to maintain a comfortable indoor temperature. It is measured in:

✓ BTU (British Thermal Units)

✓ Tons of refrigeration (1 ton = 12,000 BTU/h)

To determine the thermal load accurately, HVAC professionals use Manual J, a standard developed by the Air Conditioning Contractors of America (ACCA). This calculation considers:

✓ Building dimensions

✓ Orientation (sun exposure)

✓ Number of windows

✓ Insulation quality

A miscalculated load can lead to high energy costs and poor comfort.

- An oversized system cools too quickly but does not dehumidify properly, creating an uncomfortable indoor climate.
- An undersized system runs constantly but never reaches the desired temperature.

For commercial buildings, in addition to Manual J, professionals use:

✓ Manual D (for duct sizing)

✓ Manual S (for selecting the right HVAC unit for specific operating conditions)

Types of HVAC Systems: Which One to Choose?

Once the thermal load is calculated, the next step is selecting the most suitable HVAC system.

✓ Traditional Split Systems

- Most common in homes
- Consist of an indoor and outdoor unit
- Offer a good balance of efficiency and cost
- Ideal for single-family homes and small commercial spaces

✓ Heat Pumps

- Increasingly popular, especially in mild climates
- Provide both heating and cooling
- Use less energy than traditional fuel-based heating systems

✓ Central HVAC Systems with Ducts

- Common in large homes, offices, and retail spaces
- Ensure even temperature distribution
- Require careful duct design to minimize energy losses

✓ VRF (Variable Refrigerant Flow) Systems

- Used mainly in commercial buildings
- Offer precise temperature control for different zones
- More efficient than traditional HVAC systems

Energy Efficiency and Operating Costs

In the United States, HVAC system efficiency is measured by two key metrics:

✓ SEER (Seasonal Energy Efficiency Ratio) – For cooling systems. The higher the SEER, the greater the efficiency.

✓ HSPF (Heating Seasonal Performance Factor) – For heating systems. A higher HSPF means lower energy consumption.

As of 2023, the EPA updated the minimum efficiency standards, requiring new systems to have a SEER of at least 14-15, depending on the region.

Choosing an HVAC system with a high SEER rating can lead to significant energy savings over time.

3.2 Installation Planning: Tools, Materials, and Safety – Complete Checklist to Ensure Nothing is Overlooked

After choosing the right HVAC system, the next crucial step is installation planning. A well-organized installation ensures system efficiency, minimizes wasted time and materials, and prevents future issues.

In the United States, HVAC installation is subject to local and federal regulations. A mistake in the preparation phase can lead to safety hazards, system inefficiencies, or even building code violations. That's why it's essential to follow a detailed plan covering every aspect of the process—from choosing the right tools and materials to ensuring safety measures for both installers and the system itself.

Installation Preparation: Essential Tools

Every HVAC installation requires a specific set of tools to ensure the job is completed accurately and according to technical specifications.

- A good set of wrenches and adjustable pliers is essential for tightening fittings and securing components.
- A HVAC manifold gauge is crucial for checking refrigerant pressure, while a vacuum pump is needed to remove moisture and air from the system before charging it with refrigerant.
- Measuring tools, such as a bubble level and measuring tape, help ensure that the unit is installed perfectly level.
- A digital multimeter is essential for testing circuit continuity and diagnosing electrical connection issues.

If the installation involves air ducts, additional tools are required:

- Sheet metal cutters, metal shears, and a bending machine to shape ducts as needed.
- For refrigerant piping, a brazing set with a propane or oxy-acetylene torch is required to seal copper joints properly.

Necessary Materials for Installation

In addition to tools, having all the required materials on hand ensures the installation can be completed without delays.

- Indoor and outdoor units must be chosen based on the system specifications, ensuring they are compatible with the building layout.
- Copper tubing for the refrigerant circuit must be of the correct size and length to prevent pressure losses and energy inefficiencies.
- The ventilation system requires:
    - Galvanized sheet metal or flexible ducts
    - Insulating gaskets and sealing compounds to prevent air leaks
- Thermal insulation for refrigerant pipes helps maintain energy efficiency.
- Electrical wiring and control panels must be selected carefully to ensure safe connections between the outdoor unit, indoor unit, and thermostat.

Safety: Essential Precautions During Installation

HVAC installation involves risks related to electricity, refrigerant pressure, and construction materials. Following proper safety procedures protects installers and ensures system integrity.

- Before starting any work, power must be completely disconnected to prevent electric shocks. A voltage tester should always be used to confirm that no live currents are present in the wiring.
- When handling refrigerants, it is mandatory to wear chemical-resistant gloves and safety goggles to avoid skin contact and vapor inhalation. In case of an accidental refrigerant leak, the work area should be immediately ventilated.
- For brazing and welding, heat-resistant gloves must be worn, and work should be performed in a well-ventilated area to avoid exposure to toxic gases. The risk of fire can be minimized by using fire-resistant blankets to protect surrounding surfaces.

Complete Installation Checklist

A detailed checklist ensures that every step of the installation is completed correctly, preventing costly errors and improving work quality.

✓ Before starting installation:

- Confirm that all materials and tools are available.
- Ensure the installation site is ready for the new system.
- Check the building layout for potential obstacles that might interfere with ductwork or unit placement.

✓ During installation:

- Securely position the indoor and outdoor units.
- Install and properly connect the refrigerant piping.
- Ensure airtight sealing of all duct connections.

- Install electrical wiring and verify proper grounding.

✓ After installation:

- Test the system by running a trial cycle.
- Check refrigerant pressure and compressor operation.
- Confirm that the thermostat responds correctly.
- Verify airflow through ventilation ducts.

3.3 Key Steps for Installing a Residential HVAC System – Practical Guide with Real Examples

After carefully planning the installation and preparing all the necessary tools and materials, it's time to move on to the actual installation of a residential HVAC system.

A proper installation ensures energy efficiency, home comfort, and long system lifespan. On the other hand, errors in this phase can lead to refrigerant leaks, insufficient airflow, or compressor malfunctions. By following a structured method, complications can be avoided, ensuring the system works optimally from the first startup.

In the United States, the installation of a residential HVAC system must comply with local building codes and federal regulations, including EPA (Environmental Protection Agency) guidelines on refrigerant handling.

Positioning and Mounting the Outdoor Unit

The outdoor unit is responsible for heat dissipation and must be installed in a well-ventilated area to ensure proper thermal exchange. It is essential to place it on a stable, level base, away from obstacles like walls, fences, or trees that might block airflow.

In the United States, regulations recommend leaving at least 24 inches (about 60 cm) of clearance around the unit to allow for proper ventilation and easy maintenance access.

Once the optimal position is determined, the unit should be firmly secured on a concrete base or anti-vibration supports to reduce noise and minimize vibrations, which could damage internal components over time.

Installing the Indoor Unit and Connecting to Ductwork

The indoor unit, whether it's an evaporator for a split system or a unit connected to a duct network, must be strategically placed to ensure even air distribution throughout the home.

For centralized systems, the unit should be installed in a utility room, basement, or attic, ensuring sufficient maintenance space. If the system includes air ducts, they must be well-sealed to prevent pressure losses and thermal leaks.

If installing a ductless mini-split system, the indoor unit should be mounted at least 6 feet (1.8 meters) from the floor, away from heat sources or moisture, such as ovens, sun-exposed windows, or bathrooms.

Connecting the Refrigerant Lines and Leak Testing

Connecting the refrigerant pipes is one of the most delicate installation steps, as leaks can compromise system efficiency and cause environmental damage.

- Refrigerant lines must be carefully bent using a pipe bender to avoid kinks that could obstruct gas flow.
- After connection, a pressure test with nitrogen should be conducted to ensure there are no leaks in joints or brazed connections.

- A vacuum pump is used to remove air and moisture from the system before charging the refrigerant. This step is crucial to prevent ice formation inside the pipes and ensure proper refrigerant compression.

Electrical Connections and Thermostat Setup

The HVAC system must be properly wired to the electrical network to function safely.

- Cables must comply with NEC (National Electrical Code) standards.
- The electrical panel must be rated to support the system's power load.

Before powering the system, the wiring should be checked to ensure it is secure and free from short circuits. Once the electrical connections are made, the thermostat must be installed and properly configured.

Thermostat placement is critical for accurate temperature regulation—it should not be installed near:

- Windows
- Ventilation ducts
- Heat sources

After installation, the system must be started and tested to verify that the set temperature matches the actual room temperature.

First Startup and Final Inspection

After completing the installation, the HVAC system must be started for the first time and subjected to a series of tests to confirm proper operation.

- The first step is checking that the compressor and outdoor unit fan start correctly and that there are no abnormal vibrations.
- The airflow should be verified in all rooms, ensuring that ducts have no leaks.
- After 15-20 minutes of operation, it's recommended to:
    - Measure refrigerant pressure
    - Check temperature differences between incoming and outgoing air

If the readings match the manufacturer's specifications, the installation is successful.

3.4 Installing a Commercial HVAC System: Key Differences and Challenges – Strategies for Managing Large Spaces

After reviewing the installation process for a residential HVAC system, it's time to tackle an even more complex challenge: installing a commercial HVAC system.

Commercial HVAC systems differ significantly from residential systems in terms of size, cooling and heating capacity, air distribution, and overall complexity. These systems must be designed to handle higher thermal loads, maximize energy efficiency, and adapt to large buildings with varied climate needs across different zones.

A proper installation is essential to avoid energy waste, airflow imbalances, and comfort issues in workplaces, retail spaces, and industrial facilities. Additionally, in the United States, commercial HVAC systems must comply with stricter safety and energy efficiency standards, such as those set by the ASHRAE (American Society of Heating, Refrigerating, and Air-Conditioning Engineers) and the Department of Energy (DOE).

Key Differences Between Residential and Commercial HVAC Systems

A residential HVAC system is designed to serve a single home with a relatively limited capacity, while a commercial system must provide climate control for a much larger area, often divided into multiple zones with different climate requirements.

Main differences include:

- Capacity and Size – Commercial systems are usually modular, allowing capacity adjustments by adding or removing units as needed.
- Air Distribution – Large commercial buildings use complex ductwork systems, often with independent zones controlled by separate thermostats.
- Unit Placement – While residential units are typically placed outside the home, commercial units are often installed on rooftops (Rooftop Units, RTUs) to save space and reduce noise.
- Efficiency and Control – Commercial HVAC systems often incorporate VRF (Variable Refrigerant Flow) systems or BMS (Building Management Systems) for precise temperature regulation and optimized energy consumption.

Key Phases of a Commercial HVAC Installation

Selecting the System and Planning the Layout

Before installing a commercial HVAC system, it's crucial to determine the most suitable system for the building and carefully plan the unit and duct distribution.

- Rooftop Units (RTUs) – The most common solution for stores, restaurants, and medium-sized offices. These self-contained units simplify maintenance and maximize indoor space.
- VRF Systems – Ideal for buildings with complex climate control needs, allowing independent temperature regulation in multiple zones for greater efficiency and flexibility.

A proper layout plan includes duct design, ensuring balanced airflow to prevent areas from becoming too hot or too cold. A detailed thermal load analysis is crucial for correctly sizing the system.

Installing the Outdoor and Indoor Units

- Outdoor units must be placed on stable supports, using shock absorbers to reduce vibrations and noise.
- If installing rooftop units (RTUs), it's important to consider structural load capacity and ensure the roof can support the unit's weight.
- Indoor evaporator units are typically installed in drop ceilings or mechanical rooms, ensuring optimal air distribution without taking up operational space.

Ductwork Connection and Airflow Balancing

In commercial HVAC installations, balancing airflow is critical to ensure consistent comfort across all building zones.

- Ducts must be properly sealed to prevent pressure losses and ensure air reaches all areas without unnecessary energy waste.
- Many commercial systems use motorized dampers, which automatically regulate airflow based on zone temperature demands.
- If the system is equipped with VRF technology, electronic expansion valves must be correctly configured to ensure efficient refrigerant distribution.

Configuring the Control System and Automation

Commercial HVAC systems often integrate advanced control systems, such as BMS (Building Management Systems), which allow centralized monitoring and management of the entire HVAC network.

During installation, it is essential to:

- Test and calibrate temperature and humidity sensors.
- Verify connections to the management software.
- Set up thermostats for optimal operation.

3.5 Final Inspection and Testing: Ensuring Everything Works Perfectly – Post-Installation Checklist

After completing the HVAC system installation, it is essential to conduct a detailed inspection to ensure everything functions perfectly. Even the best system, if not properly tested before being commissioned, may experience performance issues, energy inefficiencies, or even safety hazards.

The final testing phase ensures that the system:
✓ Reaches the desired temperatures
✓ Has even airflow distribution
✓ Is free of refrigerant leaks or electrical issues

Additionally, this phase helps identify and correct any errors before the system is delivered to the client or building owner.

In the United States, HVAC system verification must comply with ASHRAE and DOE regulations to ensure the system meets efficiency and safety standards.

Electrical Connection and Thermostat Check

Before turning on the system, it is crucial to verify that all electrical connections are correct and secure. Faulty wiring can cause malfunctions, short circuits, or even fires in extreme cases.

- Inspect the electrical panel to ensure that fuses and breakers are correctly rated for the HVAC system's power load.
- Use a digital multimeter to check for voltage presence at the main terminals and thermostat connections.
- Test the thermostat in all modes (cooling, heating, and fan) to confirm that it properly communicates with the HVAC system.

If the thermostat is programmable or smart, configure it based on the client's needs to maximize energy efficiency.

Airflow Test and Duct Balancing

A properly installed HVAC system must ensure uniform airflow in all rooms or building zones.

- If some areas receive more airflow than others, the ductwork may be improperly sized, obstructed, or not properly sealed.
- Use an anemometer to measure air velocity at vents and compare it with design specifications.
- If significant differences are detected, adjust the duct dampers to balance the airflow across different zones.

Additionally, verify the proper operation of fans to ensure:

✓ They run smoothly without excessive vibrations.
✓ No unusual noises indicate mechanical issues.
✓ Fan alignment and belt tension are correctly adjusted.

Refrigerant Pressure Verification and Leak Detection

For HVAC systems using refrigerant, it is essential to ensure pressure levels are within the correct parameters for optimal performance.

- Use an HVAC manifold gauge to measure high-side and low-side pressures, comparing them with the manufacturer's recommended values.
    - If pressures are too low, there may be a refrigerant leak.
    - If pressures are too high, the system could be overcharged or obstructed.
- Check for leaks using an electronic refrigerant leak detector or the soap bubble method applied to pipe joints.
- Any identified leaks must be immediately repaired, followed by a retest of the system.

Temperature Testing and Energy Efficiency Check

After verifying that the system is electrically safe and has proper refrigerant pressure, test its heating and cooling capacity.

- Use an infrared thermometer to measure:
    - ✓ Air temperature entering the system
    - ✓ Air temperature exiting the system
- The temperature differential between the treated and incoming air should align with manufacturer specifications.

To evaluate energy efficiency:

- Measure the Seasonal Energy Efficiency Ratio (SEER) to ensure energy consumption aligns with expectations.

- Anomalies in power consumption may indicate:
  - ✓ Issues with the compressor
  - ✓ Problems with the fan motor
  - ✓ Improper system balancing

Final Post-Installation Checklist

After completing all tests, perform a final inspection to confirm that the system is ready for use. This includes:

✓ Checking for unusual noises or vibrations from the indoor or outdoor unit.
✓ Confirming air filters are correctly installed and educating the client on how and when to replace them.
✓ Testing the condensate drain system to prevent water buildup or leaks that could damage the building.
✓ Reviewing thermostat settings to ensure the system turns on and off efficiently without energy waste.
✓ Educating the client on basic maintenance procedures, including:

- Cleaning evaporator and condenser coils
- Checking and replacing air filters
- Conducting periodic HVAC system inspections

BOOK 4
Diagnosing and Solving Common HVAC System Problems

4.1 The HVAC System Won't Turn On: Causes and Solutions – Analysis and Testing for Diagnosing the Issue

After completing the HVAC system installation and initial verification, it is possible that the system does not turn on properly or does not respond to thermostat commands. This is one of the most common issues in both residential and commercial systems and can be caused by various factors, ranging from simple electrical problems to more complex failures in key components.

Diagnosing why the system won't start is the first step in resolving the issue quickly and effectively. Before intervening directly on the components, it is crucial to follow a structured analysis process to identify the root cause of the malfunction without damaging the system.

Checking the Electrical Power Supply

The first step is verifying whether the system is receiving power. If the HVAC unit is not powered, it will not turn on regardless of the thermostat settings or other components.

The first check involves inspecting the building's main electrical panel to ensure the HVAC circuit breaker has not tripped. A tripped breaker may indicate a power overload or a short circuit, which must be identified before resetting the system.

If the breaker is ON, the next step is to check the 24V transformer, which powers the HVAC control system. A faulty transformer may prevent the thermostat from communicating with the system, thus stopping it from turning on.

Another element to inspect is the HVAC fuses, either in the main electrical panel or inside the system itself. If a fuse is blown, it should be replaced with one of the same amperage, avoiding the use of an oversized fuse, which could damage the circuit.

Checking the Thermostat and Settings

If the system has power but does not turn on, the issue may be with the thermostat.

Many HVAC systems fail to start because the thermostat is incorrectly set or has a dead battery (in battery-powered models). Before checking other components, ensure that:

- The thermostat is in the correct mode (cooling or heating).
- The set temperature is lower (for cooling mode) or higher (for heating mode) than the ambient temperature.

If the thermostat is hardwired, it can be bypassed for testing by manually connecting:

- "R" and "G" terminals (to test the fan)
- "R" and "Y" terminals (to test the compressor)

If the system starts, the thermostat is faulty and needs to be replaced.

Checking the Control Relay and Contactor

If the thermostat is working, but the system still won't start, the issue may lie in the contactor or control relay.

- The contactor is an electromagnetic switch that controls power delivery to the compressor and fan. If defective or stuck, the system won't receive enough power to start.
- To test it, use a multimeter to check for continuity across the terminals.
- If the contactor does not close when the thermostat sends a start signal, it should be replaced.

Another critical component is the start relay, which sends the signal for the compressor to turn on. A faulty relay may prevent the compressor from running, even if all other components are functioning properly.

Checking the Fan Motor and Compressor

If the system powers on but does not operate correctly, there may be an issue with the fan motor or compressor.

- The fan motor should rotate freely without resistance.
- If it is stuck or making unusual noises, the bearings may need lubrication or the motor may need replacement.

The compressor, on the other hand, needs to receive the correct voltage to start.

- If the compressor attempts to start but quickly shuts off, it may indicate a faulty start capacitor, which is essential for providing the initial power boost to the motor.
- A faulty capacitor can be identified by visual inspection (if bulging or damaged) or tested with a capacitance meter to verify if the microfarad value falls within specifications.

Checking Refrigerant Pressure

If the system turns on but does not produce cold or hot air, the issue may be low refrigerant levels.

- An HVAC system with insufficient refrigerant may trigger low-pressure sensors, which prevent the compressor from starting to avoid damage.
- To verify this, an HVAC gauge manifold should be connected to the service ports to compare readings with manufacturer-recommended values.
- If the pressure is too low, there could be a refrigerant leak, which must be identified and repaired before recharging the system.

4.2 Weak or Insufficient Airflow – How to Check for Clogged Ducts, Fans, and Filters

An HVAC system may turn on correctly but fail to provide adequate airflow. This issue significantly reduces system efficiency, compromises indoor comfort, and increases energy consumption. The primary causes of insufficient airflow are blockages in the ducts, dirty or malfunctioning fans, and clogged filters.

To resolve this issue, a detailed diagnosis is necessary to identify the exact cause and implement the most appropriate solutions.

Checking Air Filters: The Primary Cause of Weak Airflow

One of the most common reasons why an HVAC system fails to provide sufficient airflow is dirty or clogged filters. The air filter's function is to trap dust, allergens, and debris, preventing them from entering the system and affecting the evaporator coil and fan operation.

A heavily clogged filter drastically reduces airflow, forcing the HVAC system to work harder, which can result in fan motor overheating and evaporator coil freezing.

To check if the filter is the problem, remove it and hold it up to the light. If light does not pass through, the filter is too dirty and must be replaced or cleaned.

In the United States, it is recommended to replace air filters every 1–3 months, depending on the filter type and indoor air quality.

Checking the Fan Blower: Dirty Blades or Motor Issues

If the filter is clean but the airflow remains weak, the next step is to inspect the indoor unit's fan, which is responsible for pushing conditioned air through the ducts.

The fan blower can have two primary issues:

- Dirt buildup on the blades, reducing the fan's ability to move air efficiently.
- Blower motor malfunction, causing low rotation speed or a complete shutdown.

To inspect the fan, open the indoor unit's access panel and check the blades. If covered in dust or debris, clean them using a damp cloth and a specialized cleaner.

If the fan does not turn on or rotates too slowly, the issue might be a faulty start capacitor, which provides the initial boost to the motor. A bad capacitor can be tested with a multimeter in capacitance mode and should be replaced if the reading falls below the manufacturer's specifications.

Inspecting Air Ducts: Blockages and Leaks

If the filter and fan are in good condition, the problem may lie within the air distribution ducts. A blockage in the ducts can significantly reduce airflow, causing uneven temperature distribution throughout the building.

The main causes of duct blockages include:

- Dust and debris accumulation, especially in systems that have not been cleaned regularly.
- Crushed or bent ductwork, reducing the available air passage area.
- Pest infestations, such as insects or small animals, which is common in buildings with exposed ductwork.

To inspect duct conditions, an HVAC inspection camera can be used—a device with a flexible probe that allows visibility inside the duct system. If blockages are found, they can be removed using an industrial vacuum, or, for severe contamination, a professional duct cleaning service may be necessary.

If no blockages are found, the issue may be an air duct leak. Leaks cause air loss before it reaches the vents, reducing system efficiency. To detect leaks, a differential pressure gauge can be used, or a smoke test can be performed at critical points in the system.

Evaporator Coil Freezing

Another possible reason for weak airflow is evaporator coil freezing. This occurs when air does not circulate properly through the coil, causing excessive temperature drop and ice formation on the heat exchange fins.

A frozen evaporator coil may result from:

- Clogged air filters, restricting airflow.
- Low refrigerant levels, causing the pressure to drop and temperatures to become too low.
- Faulty blower fan, failing to push enough air through the coil.

If the evaporator is frozen, the system should be turned off to allow the ice to melt before diagnosing and fixing the root cause.

Final Testing and Restoring Maximum Performance

After identifying and resolving the issue, it is important to test the system again to ensure that airflow has returned to normal. Using an anemometer, measure the air velocity at the vents and compare it with standard values to confirm that the system is operating correctly.

A properly functioning HVAC system with optimal airflow ensures:

✓ Efficient cooling and heating
✓ Lower energy consumption
✓ Longer system lifespan

4.3 The Compressor Won't Start: How to Test and Replace It If Necessary – Diagnostics and Repair

The compressor is the heart of the HVAC system. If it fails to start, the entire system loses its ability to cool or heat the air, making it ineffective.

When a compressor doesn't turn on, the causes can vary, including electrical issues, faulty start components, overheating, or internal motor damage. To resolve the issue, a structured diagnosis should be followed, starting from the simplest checks to the most complex tests.

Checking Power Supply and Contactor

Before focusing on the compressor itself, ensure it is receiving the correct voltage.

The first step is checking the electrical connection at the power panel. If the system is on but the compressor does not start, verify whether the dedicated breaker has tripped. If the fuse is blown, this could indicate overload or a short circuit in the compressor.

The contactor is the component that supplies power to the compressor when the thermostat sends the start command. If the contactor is faulty, the compressor will not receive power.

To test the contactor, use a multimeter to check the voltage at its terminals when the thermostat calls for cooling. If the contactor does not close the circuit, it must be replaced.

Testing the Start Capacitor and Start Relay

If the contactor is working properly but the compressor still won't start, the issue may be with the start capacitor. This component provides an initial boost to the compressor motor.

A faulty capacitor can often be identified visually, showing bulging or leaking dielectric fluid. However, the most accurate way to test it is with a multimeter in capacitance mode. If the measured value is lower than the rating on the capacitor, it should be replaced.

Additionally, the start relay may be defective, preventing the compressor from receiving the start signal. If the relay does not send power to the capacitor, the compressor will not receive the charge needed to start.

Checking the Compressor's Thermal Conditions

A compressor that won't start may be in thermal protection mode. If it overheated due to extended operation, low refrigerant levels, or poor ventilation, it may require 30–60 minutes to cool down before attempting to restart.

To check the compressor's temperature, use an infrared thermometer. If the temperature exceeds 140°F (60°C), it indicates the compressor has been running under critical conditions.

If the compressor frequently overheats, there may be:

- A refrigerant leak, causing improper system pressure.
- Low circuit pressure, preventing proper heat dissipation.
- A condenser fan malfunction, failing to remove excess heat.

Testing the Compressor Windings for Continuity

If all start components are functioning but the compressor still won't start, the internal motor must be tested.

The compressor motor consists of three windings:

- Common (C)
- Start (S)
- Run (R)

A continuity test with a multimeter determines whether the motor is functional or has an internal short circuit.

To perform the test:

- Disconnect the compressor from the power supply.
- Remove the cover from the connection box.
- Set the multimeter to resistance mode (Ohms).
- Measure resistance between C-S, C-R, and S-R terminals.

If the resistance between Common and Start is higher than between Common and Run, the compressor is in good condition. If there is no continuity, the windings are burned out, and the compressor must be replaced.

Replacing the Compressor: When Is It Necessary?

If the compressor is damaged beyond repair, it must be replaced. Compressor replacement is a complex procedure requiring specialized tools and safe refrigerant handling.

To replace a compressor, the following steps must be followed:

1. Recover the refrigerant using a recovery pump to prevent environmental release.
2. Disconnect the compressor electrically and remove the refrigerant line connections.
3. Install the new compressor, ensuring compatibility with the refrigerant type in the system.
4. Vacuum the refrigerant circuit to remove moisture and air.
5. Recharge the refrigerant following the manufacturer's specifications.
6. Perform a final test to confirm the new compressor is working correctly.

Compressor replacement requires specific certifications for handling refrigerants, as mandated by the EPA (Environmental Protection Agency) in the United States.

4.4 Refrigerant Leaks: How to Detect and Repair Them Properly – Practical Techniques with Specialized Tools

Refrigerant is the essential fluid for the proper operation of an HVAC system. If the refrigerant level drops below the necessary amount, the entire system loses efficiency, the compressor operates under stress, and cooling or heating becomes ineffective. A refrigerant leak can lead to serious issues, such as evaporator freezing, excessive energy consumption, and, in the worst case, permanent compressor damage.

Detecting and repairing a leak promptly is crucial to prevent system damage and reduce refrigerant waste, which can also have a significant environmental impact. In the United States, refrigerant management is regulated by the EPA (Environmental Protection Agency), requiring certified tools for leak detection and system recharging.

Signs of a Refrigerant Leak

An HVAC system with a refrigerant leak displays specific symptoms. The first sign is a reduced cooling or heating capacity. If the system takes longer than usual to reach the desired temperature, there may be a gas leak.

Another indication is the formation of ice on the evaporator coil. When the refrigerant level is too low, the system pressure drops, and the evaporator temperature falls below freezing, causing moisture in the air to freeze. This leads to further airflow obstruction and a drastic drop in performance.

In systems with a significant leak, a hissing sound or slight air leakage noise may be heard near the refrigerant pipes or compressor fittings. Some units may also emit a burnt oil smell, caused by refrigerant mixing with compressor lubricant.

Techniques for Detecting Refrigerant Leaks

Once a leak is suspected, it must be confirmed using specific tools.

One of the simplest methods is the soap bubble test, where a soapy solution is applied to fittings and pipes to check for bubbles indicating a leak. While cost-effective, this method may not be effective for detecting small leaks.

Using an electronic refrigerant leak detector is the most reliable solution. These devices can detect even minimal concentrations of gas in the air and can identify leaks in hard-to-reach areas.

Another advanced method is UV dye inspection. A fluorescent dye is injected into the refrigerant circuit, and after a few hours of operation, a UV light is used to identify leakage points.

If the leak is difficult to detect, a nitrogen pressure test can be performed. The system is evacuated and filled with high-pressure nitrogen. If the pressure drops rapidly, this indicates a leak somewhere in the circuit. This method is especially useful for internal leaks in condensers and evaporators.

Repairing Refrigerant Leaks

Once the leak is located, the appropriate repair technique must be selected. If the leak is in a threaded fitting, tightening the connection and applying a refrigerant sealant may be sufficient.

If the leak is in a copper pipe, silver brazing can be used to create a strong, airtight seal. Before brazing, the system must be fully evacuated and filled with nitrogen to prevent carbon buildup inside the pipes.

If the leak is in a complex component like the evaporator or condenser, it may be more cost-effective to replace the entire component rather than attempt a repair.

After the repair, the system must be re-pressurized with nitrogen and tested to confirm that the leak has been completely sealed before recharging the refrigerant.

Refrigerant Recharge and Final Testing

After repairing the leak, the system must be recharged with the exact amount of refrigerant specified by the manufacturer.

A refrigerant scale is essential to ensure that the system receives the correct gas volume. Too much refrigerant can be just as harmful as too little, causing abnormal pressure levels and reducing system efficiency.

Once the recharge is complete, the system must be tested in both cooling and heating modes to verify that pressure levels are within the optimal range and that the air output matches the expected temperature.

The final check involves ensuring there are no new leaks. Even a minor overlooked leak can eventually decrease system performance and lead to another refrigerant drop.

Impact of Refrigerant Leaks on the System and the Environment

Beyond reducing system efficiency, a refrigerant leak can have a major environmental impact. Certain refrigerants, like R-22, have been phased out in the United States due to their contribution to ozone depletion and climate change.

The transition to eco-friendlier refrigerants, such as R-32 and R-410A, has mitigated this impact, but leaks must still be carefully managed. Releasing refrigerant into the atmosphere is illegal and can result in fines and penalties imposed by the EPA.

Every certified HVAC technician must comply with EPA Section 608 regulations, which require the proper recovery and disposal of refrigerants during maintenance and repair operations.

A properly sealed and well-maintained HVAC system ensures:

- Optimal performance
- Lower energy consumption
- Longer lifespan for the compressor and other components

4.5 Unusual Noises and Vibrations: When to Be Concerned and How to Fix Them – Identification and Solutions for Abnormal Sounds

A properly functioning HVAC system should operate relatively quietly, producing only a low humming sound from the compressor and fan. However, if the system starts emitting unusual noises or excessive vibrations, there could be an underlying issue requiring immediate attention.

Abnormal sounds are not just annoying—they may indicate worn-out components, loose screws, electrical issues, or even structural damage to the system. Ignoring these warning signs can lead to more severe and costly failures. Correctly identifying the noise source allows for timely intervention to prevent further issues, ensuring efficient and safe HVAC operation.

Types of Noises and Their Causes

Each type of noise coming from an HVAC system has a specific underlying cause. Understanding the difference between humming, clicking, whistling, or banging sounds is crucial for pinpointing the issue and resolving it efficiently.

- A constant humming noise may indicate that the fan motor or electrical transformer is overloaded. If the humming sound is accompanied by motor overheating, there could be an issue with the start capacitor or the fan imbalance.
- A regular clicking sound is often caused by debris or foreign objects lodged in the fan blades or condenser fins. If the noise occurs only during startup, it could be a faulty relay struggling to close the circuit.
- A high-pitched whistling noise is a warning sign of a potential refrigerant leak. If the sound comes from the evaporator or refrigerant lines, it is crucial to check gas pressure immediately to prevent compressor damage.
- A loud banging or metallic noise may indicate loose screws, improperly secured supports, or a more serious compressor issue. If the noise originates from the outdoor unit, the compressor may be internally damaged and may require replacement.

Identifying the Source of the Noise

Before taking corrective action, it is essential to locate the source of the noise to determine which component is responsible. The first step is to carefully listen to whether the sound is coming from the indoor unit, outdoor unit, or air ducts.

- If the noise comes from the indoor unit, check the fan, motor, and electrical panel. Loose screws or damaged mounts can amplify vibrations and create metallic noises.
- If the sound originates from the outdoor unit, the problem could be in the compressor, condenser fan, or refrigerant pipes. A hissing noise may indicate a gas leak, while a loud hum could mean the compressor is under stress.
- If the noise is coming from the air ducts, there could be an obstruction creating air turbulence, or loose ductwork sections vibrating due to improper installation.

Troubleshooting and Fixing Common Issues

Once the noise source is identified, corrective actions can be taken.

- If the issue involves loose mounts or vibrating components, tightening screws and bolts with a screwdriver or wrench can stabilize moving parts. In the case of the compressor, using anti-vibration mounts helps reduce noise and minimize motor wear.
- If the noise comes from the fan, turn off the system and inspect the blades for debris or signs of wear. If the blades are unbalanced, they may be bent or loose and should be adjusted or replaced.
- If a constant hissing sound is detected, check refrigerant pressure and look for leaks along the piping. If low pressure is found, identify and repair the leak before recharging the system.
- A loud noise from the compressor may indicate loose internal supports or motor damage. If the compressor makes a metallic sound when attempting to start, the start capacitor or the compressor itself may need replacement.

When Noise Indicates a Serious Problem

Some abnormal sounds are signs of severe issues that require immediate action.

- If a loud bang is heard when the compressor starts, there could be an internal piston problem, potentially leading to complete failure.
- A gradually increasing hum may indicate that the fan motor is close to seizing up. If the noise gets louder over time, replacing the motor before it fails completely is recommended.
- A strong vibration noise from air ducts may suggest that the ducts were installed incorrectly, or sections have collapsed, reducing airflow efficiency and making the system less effective.
- If the noise is caused by cavitation in the refrigerant circuit, the compressor may be running with an insufficient gas charge. This problem must be resolved immediately to prevent irreversible damage.

Final Testing and Restoring Silent Operation

After identifying and resolving the noise issue, it is important to test the system to confirm that the sound has been eliminated. The system should be turned on and left running for a few minutes, while listening for any remaining vibrations.

Using anti-vibration pads and acoustic insulation around the outdoor unit can help reduce future noise. Regular maintenance, including cleaning filters and checking electrical connections, also helps prevent unwanted noises and ensures quieter operation over time.

BOOK 5
Preventive Maintenance and Energy Efficiency Optimization

5.1 Cleaning and Replacing Air Filters: When and How to Do It Properly – Periodic Maintenance Checklist

Air filters are an essential component of any HVAC system. Their primary role is to trap dust, debris, allergens, and other impurities, preventing these particles from entering the system and compromising its performance. A clean filter improves energy efficiency, reduces operating costs, and extends the lifespan of the system's internal components.

If a filter becomes dirty or clogged, airflow is significantly reduced, forcing the fan to work harder to push air through the system. Additionally, the compressor may overheat due to the increased workload. This not only raises energy consumption but can also lead to expensive repairs over time.

In the United States, HVAC manufacturers recommend replacing or cleaning filters every 1–3 months, depending on the type of filter and environmental conditions in the home or commercial building.

When to Replace or Clean Air Filters?

The right time to replace or clean an air filter depends on several factors, including the type of filter used, the amount of dust in the environment, and the frequency of HVAC usage.

- Disposable fiberglass filters should be replaced every month. They are affordable but less effective at capturing smaller particles.
- Pleated filters offer better filtration capacity and can last up to three months before needing replacement.
- If using a washable electrostatic or HEPA filter, it must be cleaned regularly to maintain optimal performance. A washable electrostatic filter can be rinsed with warm water and left to dry completely before reinstalling.

A clear sign that a filter needs replacement is reduced airflow from ventilation registers. If the system struggles to maintain the desired temperature or if the air output feels weaker than usual, the filter is likely clogged.

Indoor air quality is another indicator. If dust buildup increases, unpleasant odors persist, or humidity levels rise, the filter may no longer be effectively trapping impurities.

How to Properly Replace an Air Filter

Replacing an air filter is one of the simplest maintenance tasks, but it must be done correctly to ensure optimal HVAC performance.

1. Turn off the HVAC system to prevent dust and debris from circulating while replacing the filter.
2. Locate the filter compartment, usually found in the return air duct near the indoor unit.
3. Remove the old filter and check its size to ensure a proper replacement.
4. Install the new filter, making sure to follow the airflow direction marked by an arrow on the filter's edge.

If the system uses a washable filter, it should be cleaned with warm water and, if needed, a mild detergent to remove stubborn debris. After washing, the filter must dry completely before reinstalling to prevent mold formation.

After completing the replacement, record the date to remember when the next check is due.

Benefits of Regular Filter Maintenance

Keeping filters clean provides numerous advantages. An HVAC system with properly maintained filters operates more efficiently, reducing energy consumption and ensuring a comfortable indoor environment.

- Optimal airflow allows the system to evenly distribute heating or cooling, avoiding hot and cold spots in the house.
- Clean filters reduce dirt buildup on evaporator coils and condenser units, preventing performance issues over time.
- Better indoor air quality—filters trap dust, pollen, bacteria, and other impurities, reducing allergy and respiratory risks, which is especially important for asthma sufferers.
- Less strain on the fan motor—a clogged filter forces the fan to work harder, consuming more energy and increasing the risk of overheating.

Periodic Maintenance Checklist for HVAC Filters

To ensure air filters remain in optimal condition, following a regular maintenance checklist is essential.

- Every month: Visually inspect the filter to determine if cleaning or replacement is needed. If the home is in a dusty or high-pollen area (e.g., Southwestern U.S.), more frequent checks may be necessary.
- Every three months: Perform a standard filter replacement or a deep cleaning for washable filters.
- At least once a year: Conduct a full HVAC system check, including the fan, evaporator, and compressor, along with filter replacement.
- If using a HEPA or electrostatic filter, verify that the filtration system functions properly and replace worn-out components if necessary.

5.2 Inspection and Maintenance of the Compressor and Fans – Techniques to Extend Their Lifespan

The compressor and fans are two of the most critical components of an HVAC system. The compressor is responsible for circulating refrigerant through the system, while the fans ensure proper heat exchange with the surrounding environment. If either of these components malfunctions, the entire HVAC system loses efficiency, leading to higher energy consumption and reduced cooling or heating capacity.

Regular maintenance of the compressor and fans is essential to prevent premature failures and extend the system's lifespan. In the United States, preventive maintenance is recommended by major HVAC manufacturers to avoid costly repairs and ensure maximum energy efficiency.

Compressor Inspection: Signs of Malfunction

The compressor operates under high pressure and temperature. If not properly maintained, it can overheat, consume more energy, or even suffer irreversible failure.

- The first check involves listening to the noise levels. A healthy compressor emits only a low humming sound while running.
- If you hear metallic noises, clicking sounds, or loud knocking, there may be internal issues, such as damaged pistons or worn-out mounts.
- Compressor temperature is another critical indicator. If too hot to touch, it may be overloaded or running with insufficient refrigerant.
- Using an infrared thermometer, you can check whether the temperature falls within the normal operating range specified by the manufacturer.
- Dirt and debris accumulation on the compressor surface reduces its ability to dissipate heat, leading to overheating. Regular cleaning with a dry cloth or compressed air helps maintain optimal temperature levels.

Compressor Maintenance: Preventing Overheating

If the compressor frequently overheats, it's crucial to address the issue before it results in permanent damage.

- The first step is to check refrigerant pressure using an HVAC pressure gauge.
    - High pressure may indicate a blockage in the system or excess refrigerant.
    - Low pressure may signal a leak, which must be identified and repaired.
- Inspect the start capacitor—a faulty capacitor prevents the compressor from starting properly, causing short cycling that increases motor stress.
    - Using a multimeter in capacitance mode, you can check if the capacitor is providing the correct charge.
- Inspect electrical wiring for loose connections or damaged cables.
    - If wires show signs of burning or corrosion, they should be replaced immediately to prevent short circuits.

Condenser Fan Inspection and Cleaning

The condenser fan plays a crucial role in dissipating heat generated by the compressor. If it fails, the compressor must work harder to maintain system temperature, leading to higher energy costs and potential overheating risks.

- The first sign of a fan problem is reduced or uneven airflow from the outdoor unit.
- If the fan spins slowly or produces excessive vibrations, it may need cleaning or realignment.
- Dust, leaves, and debris on the fan blades can reduce airflow efficiency.
    - Cleaning the blades regularly with a damp cloth or soft brush prevents cooling issues.
- Unusual noises from the fan could indicate worn-out motor bearings.
    - Damaged bearings cause excessive friction, which may lead to motor failure.
    - Applying a special lubricant may help, but if the damage is severe, the motor should be replaced.

Evaporator Fan Balancing and Adjustment

The evaporator fan is responsible for pushing cooled or heated air through the duct system. If it malfunctions, airflow may become weak or unbalanced.

- The first step is to check that the blades are securely attached and balanced.
    - A misaligned fan can cause excessive vibration, leading to premature wear on the motor and electrical connections.
- If the fan blades are bent or loose, they should be replaced to ensure even airflow.
- Adjusting motor speed is also important.
    - If the fan spins too slowly, the run capacitor may need adjustment, or the wiring should be checked.
- If the evaporator fan doesn't start, test the control relay or thermostat.
    - A faulty relay may block the fan's activation signal, preventing the system from working properly.

Final Check and System Optimization

After inspecting and cleaning the compressor and fans, it's crucial to run a full system test to ensure everything functions correctly.

- Turn on the HVAC system and monitor the pressure, temperature, and airflow parameters.
- Using diagnostic tools such as:
    - An anemometer to measure airflow.
    - A multimeter to test electrical components.
    - An infrared thermometer to monitor compressor temperature.

Maintaining the compressor and fans in good condition ensures an efficient HVAC system, reduces energy costs, and prevents expensive breakdowns. With regular maintenance, the entire HVAC unit will operate more quietly, reliably, and with lower utility bills.

5.3 Thermostat Adjustment and Strategies to Reduce Energy Consumption – Optimizing Temperature Without Waste

Thermostat regulation is one of the most important factors in optimizing the energy consumption of an HVAC system. Incorrect settings can lead to significant energy waste, higher utility bills, and reduced system efficiency.

In the United States, where heating and air conditioning are widely used throughout the year, knowing how to properly set the thermostat can mean the difference between efficient operation and unnecessary energy consumption. Strategies such as programmable thermostats, adjusting temperatures based on schedules, and monitoring system performance help maintain comfort while minimizing waste.

Ideal Thermostat Settings for Each Season

The U.S. Department of Energy (DOE) recommends specific temperature settings to maximize efficiency.

- In winter, the recommended heating temperature is 68°F (20°C) when at home, and it can be lowered by 7-10°F at night or when the house is unoccupied.
- In summer, the recommended cooling temperature is 78°F (25°C) when at home, and it should be increased if the house is empty.

Adjusting the temperature by just a few degrees based on time of day can reduce energy consumption without affecting comfort. For every degree lower in winter or higher in summer, it is possible to save up to 10% on energy bills.

Benefits of Programmable and Smart Thermostats

Using a programmable or smart thermostat is one of the most effective strategies to reduce energy consumption while maintaining comfort.

- A programmable thermostat allows users to set precise schedules for turning the system on and off, preventing unnecessary heating or cooling when the home is empty.
- Smart thermostats offer even more advanced features, such as:
    - Learning user habits
    - Remote control via mobile apps
    - Automatic adjustments based on weather conditions

Smart thermostats can also integrate with home automation systems and voice assistants like Alexa, Google Assistant, and Apple HomeKit, making it possible to control indoor climate settings via voice command or remotely.

Avoiding Waste with Smart Temperature Adjustments

A common mistake is setting extreme temperatures in summer (too low) or in winter (too high), thinking that the HVAC system will heat or cool faster.

- In reality, the HVAC system always operates at the same speed, regardless of the temperature setting.
- Extreme settings result only in higher energy consumption without improving heating or cooling times.

Other energy-saving practices include:

- Using the "fan-only" mode during the cooler hours of the day to circulate air without activating the compressor.

- In winter, lowering the heating temperature slightly and using warm clothing and blankets instead.
- In summer, keeping blinds and curtains closed during the hottest hours to prevent solar heat from increasing indoor temperatures, reducing the air conditioner's workload.

Balancing Temperatures Between Rooms

An HVAC system works more efficiently when airflow is evenly distributed. To achieve proper thermal balance, it's important to:

- Ensure that all air vents are open and not blocked by furniture or curtains.
- Address uneven temperatures between floors:
    - In winter, the upper floor is often warmer due to rising heat.
    - In summer, the upper floor can feel colder than the lower levels.
    - To fix this, adjust air vents or install a zoned HVAC system, allowing different temperatures in different parts of the house.
- Ceiling fans are an excellent tool for improving heat or cooling distribution:
    - In winter, set the fan to rotate clockwise to push warm air downward.
    - In summer, reverse the direction to create airflow that helps cool rooms more effectively.

Monitoring Energy Consumption and Thermostat Maintenance

Regularly monitoring energy usage can identify system inefficiencies. If the utility bill increases without an obvious reason, the HVAC system may be inefficiently regulated or require maintenance.

- Cleaning the thermostat at least once a year prevents dust accumulation from affecting temperature readings and causing inaccurate adjustments.
- Checking the wiring ensures there are no loose connections that could cause malfunctions or frequent on/off cycles.

- If the thermostat has an internal battery, it should be replaced periodically to avoid losing programmed settings.

A well-regulated HVAC system reduces energy consumption, extends the lifespan of components, and ensures consistent comfort throughout the year. Investing in an efficient thermostat and adopting energy-saving strategies leads to significant long-term cost savings.

5.4 Improving Indoor Air Quality – Strategies and Solutions for Healthier Air

An HVAC system does more than just heat or cool indoor spaces; it also plays a crucial role in maintaining optimal indoor air quality. Breathing clean, well-filtered air is essential for the health and well-being of those living or working indoors, especially individuals with allergies, asthma, or other respiratory conditions.

In the United States, where people spend an average of 90% of their time indoors, ensuring high air quality is a top priority. Poor indoor air, filled with pollutants and stale air, can cause fatigue, respiratory problems, eye irritation, and reduced concentration. Improving air quality requires a combination of proper HVAC maintenance, air purification techniques, and effective ventilation strategies.

The Role of Air Filters in Indoor Air Quality

One of the first aspects to check for improving air quality is the condition of HVAC filters. These filters capture dust, bacteria, pollen, and other airborne particles, preventing them from circulating indoors.

Not all filters are the same:

- Standard fiberglass filters provide basic protection but do not trap smaller particles effectively.
- High-Efficiency Particulate Air (HEPA) filters can capture up to 99.97% of particles as small as 0.3 microns, including viruses and bacteria.

- Activated carbon filters are particularly useful for absorbing odors, chemical fumes, and volatile organic compounds (VOCs)—substances emitted by paints, furniture, and cleaning products.

Replacing HVAC filters regularly, typically every 1-3 months, is essential to ensure the system continues filtering air effectively without reducing airflow.

Humidity and Air Quality: The Role of Humidifiers and Dehumidifiers

Humidity levels significantly impact indoor air quality.

- Dry air can cause respiratory irritation, dry skin, and increased dust accumulation.
- Excess humidity promotes the growth of mold, bacteria, and dust mites.

The ideal indoor humidity level for comfort and health is between 30% and 50%.

- If indoor air is too dry, especially during winter when heating is on, using a humidifier helps maintain a comfortable environment.
- In areas with high humidity, using a dehumidifier prevents moisture buildup on walls and reduces mold growth.

Some modern HVAC systems include integrated humidity control features, allowing for automatic regulation of optimal humidity levels.

Ventilation and Air Exchange: Why It's Essential

A properly functioning HVAC system must ensure adequate air exchange. Modern buildings are often highly insulated to reduce energy consumption, but this can lead to the accumulation of carbon dioxide and pollutants due to insufficient ventilation.

One of the best solutions is using mechanical ventilation systems (VMC), which bring in fresh outdoor air while removing stale indoor air—all without significant heat loss.

When the weather allows, opening windows for at least 10-15 minutes daily can help reduce pollutant concentrations and improve indoor comfort. However, in areas with high outdoor pollution or pollen, it is better to rely on filtered ventilation systems.

Air Purifiers: When Are They Necessary?

In cases where an HVAC system alone is not enough to maintain optimal air quality, using standalone air purifiers can be a highly effective solution.

- The best air purifiers use HEPA and activated carbon filters to eliminate microscopic particles and reduce odors.
- Some advanced models include UV light technology to neutralize bacteria and viruses, increasing safety indoors.

Using an air purifier is particularly recommended for:

- Homes with pets
- People with allergies or respiratory issues
- Areas with high air pollution

Controlling Indoor Pollution Sources

Beyond improving filtration and ventilation, it is essential to reduce indoor pollution sources.

- Avoid scented candles, aerosol sprays, and harsh chemical cleaners, as they contribute to volatile organic compounds (VOCs) in the air.
- Using low-emission paints and natural materials for furniture reduces the accumulation of harmful chemicals indoors.

Regular HVAC maintenance is also crucial to prevent mold and bacteria buildup within the air handling units. Ductwork should be inspected periodically to prevent dust accumulation and the spread of allergens throughout the home.

Another effective approach is using indoor plants that naturally filter air pollutants. Some of the best plants for air purification include:

- Sansevieria (Snake Plant)
- Dracaena
- Ficus

These plants are especially effective at absorbing common pollutants such as formaldehyde and benzene.

5.5 Planning Professional HVAC Maintenance: When to Call an Expert – How to Identify Critical Malfunctions

A well-maintained HVAC system can operate efficiently for years without major breakdowns. However, even with regular owner maintenance, there are situations where professional intervention is necessary. Recognizing early warning signs of malfunctions can prevent serious damage, avoiding expensive repairs or premature system replacement.

In the United States, HVAC safety regulations require that certain operations, such as handling refrigerant or diagnosing advanced electrical issues, be performed only by EPA Section 608-certified technicians. Knowing when to call an expert not only ensures system safety but also helps maintain the manufacturer's warranty.

Signs of Malfunctions That Require Professional Service

Some HVAC issues can be resolved with basic maintenance, such as cleaning filters or adjusting the thermostat. However, certain warning signs indicate the need for a thorough inspection by a certified technician.

- If the system does not turn on or shuts off unexpectedly, there may be an issue with the electrical wiring, thermostat, or compressor control panel. A technician can perform advanced diagnostics to identify the faulty component.
- If there is a refrigerant leak, calling a professional is essential. Handling refrigerant is regulated by law and requires specialized equipment to detect and seal leaks without releasing harmful gases into the environment.
- If the system makes unusual noises, such as ticking, metallic banging, or loud buzzing, it could indicate a problem with the compressor motor, fan, or internal bearings. A technician can disassemble and inspect the components for wear and replace any damaged parts.
- If the air output is not at the correct temperature, the issue may be due to incorrect refrigerant pressure, a frozen evaporator, or a faulty expansion valve. A professional can test these components and rebalance the system.

Recommended Frequency for Professional Maintenance

Even if the HVAC system is functioning properly, scheduling professional maintenance at least once a year is recommended.

- Certified HVAC technicians perform in-depth tests that go beyond routine cleaning and filter replacement.
- An annual inspection includes:
    - Checking refrigerant pressure
    - Testing the compressor
    - Inspecting the electrical circuit for loose connections or worn-out components

In the United States, many HVAC service providers offer annual maintenance plans at a fixed cost, ensuring the system is checked before critical seasons, such as winter for heating and summer for cooling.

Regular maintenance not only extends the lifespan of the HVAC system but also prevents warranty invalidation. Some manufacturers require proof of annual maintenance to cover repairs under warranty.

Advanced Diagnostic Tests Performed by Technicians

During a professional inspection, technicians use advanced diagnostic tools to detect potential problems before they become serious.

- A digital multimeter is used to test electrical continuity and verify proper operation of the compressor, fans, and power transformer.
- An HVAC pressure gauge measures refrigerant pressure to determine whether the system needs a recharge or if there are leaks in the circuit.
- An infrared thermometer helps detect overheating in electrical components or check the air temperature at different system points to ensure optimal performance.
- Some technicians use thermal imaging cameras to identify heat loss in buildings or detect airflow distribution issues in ductwork.

When the Issue is Too Severe: Repair or Replace the System?

If the HVAC system is over 10-15 years old, replacing it may be more cost-effective than continuing to pay for frequent repairs. Older systems tend to be less energy-efficient and develop recurring problems, increasing maintenance costs.

A qualified technician can perform a cost-benefit analysis to determine whether repairing the issue or investing in a new system is the better option.

- New high-efficiency HVAC systems can reduce energy consumption by up to 30%, allowing homeowners to recoup their investment within a few years through lower operational costs.
- If the compressor fails and is out of warranty, it is often more economical to replace the entire outdoor unit rather than repair a single component.
- If the issue is limited to a fan motor or capacitor, the repair is usually more cost-effective than replacing the entire system.

How to Choose a Reliable HVAC Technician

In the United States, it is essential to ensure that an HVAC technician is NATE-certified (North American Technician Excellence) and holds the required state licenses.

- A qualified technician should provide a detailed estimate before performing any repairs and clearly explain the issue before proceeding.
- It is always advisable to get multiple estimates to compare costs and services.

Many HVAC service providers offer scheduled maintenance plans, which include annual inspections and discounts on repairs. These contracts are an excellent way to keep the system in good condition and reduce the risk of unexpected breakdowns.

BOOK 6
Tools and Techniques for Advanced Diagnostics

6.1 How to Use a Multimeter to Test Electrical Components

A multimeter is an essential tool for diagnosing and troubleshooting problems in HVAC systems. Electrical failures can prevent the system from turning on, cause intermittent malfunctions, or even damage internal components. Knowing how to properly use a multimeter allows technicians to test continuity, voltage, and resistance, quickly identifying any electrical issues.

In the United States, electrical safety regulations require specific precautions when working with high-voltage circuits. Wearing insulated gloves and ensuring that the system is disconnected from the power supply before performing any tests is essential to avoid electrocution risks.

Structure and Function of a Multimeter

A multimeter consists of:

- A digital or analog display
- A function selection dial
- Two test probes (red for positive and black for negative)

Depending on the selected mode, it can measure:

- Voltage (V)
- Current (A)
- Resistance (Ω)

More advanced multimeters offer additional functions, such as capacitor testing and frequency measurement, which are useful for more detailed diagnostics.

Measuring Voltage in an HVAC System

The first test performed on an HVAC system is checking the power supply voltage. If the unit does not turn on, there may be a disruption in the electrical current flow.

To measure voltage:

1. Set the multimeter to AC Voltage mode (V~) if testing alternating current from the electrical grid.
2. Insert the black probe into the COM terminal and the red probe into the VΩ terminal.
3. Place the probes on the power input terminals of the HVAC unit.
4. Read the value on the display:
    - Residential HVAC systems in the U.S. should show approximately 120V for low-voltage circuits and 240V for high-voltage units.

If the value is zero or lower than normal, there may be a wiring issue, a blown fuse, or an electrical line interruption.

Testing Continuity and Resistance

Another critical test is checking the continuity of wires and electrical components to detect circuit breaks.

To perform a continuity test:

1. Set the multimeter to continuity mode (indicated by a sound wave symbol or Ω for resistance).
2. Turn off the HVAC system and disconnect it from power.
3. Connect the probes to both ends of the component to be tested (e.g., fuses, thermostat, relays).
4. If the multimeter beeps or shows a low resistance value (Ohms), the circuit is continuous.
5. If the reading is infinite or the multimeter does not beep, the circuit is open, indicating a faulty component.

This test is especially useful for checking protection fuses, which are often responsible for sudden failures in HVAC systems.

Testing a Start Capacitor

The start capacitor provides the initial energy boost needed to start the compressor and fan motor. If the capacitor is faulty, the motor may not start or struggle to run.

To test a capacitor:

1. Set the multimeter to capacitance mode (µF).
2. Disconnect the capacitor from the system and discharge any residual energy using an insulated screwdriver.
3. Connect the probes to the capacitor terminals.
4. Compare the reading with the rated value on the capacitor.
    - If the measured value is significantly lower, the capacitor needs replacement.
    - If the capacitor is completely faulty, the multimeter may show no reading, indicating an open circuit.

Identifying HVAC Circuit Problems with a Multimeter

Using a multimeter allows technicians to diagnose various electrical issues in HVAC systems. If the compressor does not start, the thermostat is unresponsive, or the fans do not spin, voltage, continuity, and capacitance tests help identify the root cause.

- If the thermostat does not send signals to the system, check the voltage at its terminals to determine if the issue is with wiring or the thermostat itself.

- If the compressor does not receive power, the problem could be in the contactor. A multimeter test on the contactor terminals will determine if the component is still functional or needs replacement.
- In systems with thermal protection, a high resistance reading on temperature sensors may indicate that a safety switch is preventing the system from starting to protect the compressor from overheating.

6.2 How to Measure Refrigerant Pressure with an HVAC Manifold Gauge

Measuring refrigerant pressure is an essential step in HVAC system diagnostics and maintenance. Abnormal pressure levels can indicate various issues, such as refrigerant leaks, system blockages, or compressor malfunctions. Using an HVAC manifold gauge correctly helps determine whether the system is operating with the correct refrigerant charge and if pressure imbalances are affecting efficiency.

In the United States, EPA Section 608 regulations mandate that only certified technicians handle and recharge refrigerants. However, pressure measurement can also be performed by non-certified professionals for diagnostic purposes, as long as no refrigerant is released into the atmosphere.

Types of Pressures in HVAC Systems and Their Meaning

In an HVAC circuit, two main pressures are measured:

- High-Side Pressure: The pressure of the refrigerant after compression, typically in the condenser.
- Low-Side Pressure: The pressure of the refrigerant before compression, generally in the evaporator.

Abnormal values in either pressure indicate different system issues:

- Excessively high pressure may suggest a blocked condenser, an overcharged system, or a faulty condenser fan.
- Excessively low pressure may indicate a refrigerant leak or a restriction in the system's capillary tubes.

Structure of an HVAC Manifold Gauge and How to Use It

An HVAC manifold gauge consists of:

- Two dials with indicator needles
- Three flexible hoses for system connection

The dials are color-coded for easy use:

- Blue = Measures low pressure (connects to the suction line)
- Red = Measures high pressure (connects to the discharge line)
- Yellow = Used for refrigerant charging or recovery

How to Connect the Manifold Gauge:

1. Turn off the HVAC system and let it cool down for a few minutes.
2. Attach the hoses to the system's service valves (Blue = Low Side, Red = High Side).
3. Slowly open the valve to allow pressure readings.

Reading Pressure and Diagnosing the System

Once the manifold gauge is connected, the next step is interpreting the readings. Each refrigerant type has specific operating pressures, so it's essential to consult reference tables for the specific refrigerant used in the system.

- If low-side pressure is too low, the system may have a refrigerant leak, a restricted expansion valve, or a faulty compressor.
- If high-side pressure is too high, it could indicate a blocked condenser, excess refrigerant, or poor ventilation around the outdoor unit.

In a properly functioning system:

- Low-side pressure for R-410A refrigerant should be between 60-80 psi under normal cooling conditions.
- High-side pressure varies more, typically ranging between 200-300 psi, depending on outdoor temperature.

Advanced Testing: Pressure Differences and Superheat Measurement

Beyond direct pressure readings, HVAC technicians calculate two critical parameters:

- Superheat: The temperature difference between the refrigerant leaving the evaporator and the evaporation temperature corresponding to measured pressure.
    - Too low → The system may be overcharged with refrigerant.
    - Too high → Indicates a possible leak or restricted flow in the circuit.
- Subcooling: Measures how much the refrigerant is cooled before entering the evaporator.
    - Abnormal values could suggest an inefficient condenser or faulty expansion valve.

Superheat and subcooling are measured using the manifold gauge along with a temperature probe to obtain the actual pipe temperature.

Common Mistakes in Pressure Measurement and How to Avoid Them

Common mistakes when measuring refrigerant pressure include:

- Not checking ambient temperature before measurement
    - Refrigerant pressure varies with outdoor temperature, so reference values must be adjusted accordingly.
- Opening the manifold valves too quickly
    - This can cause a sudden refrigerant release, harming both the environment and technician safety.
- Inconsistent gauge readings
    - May be caused by loose hose connections or air trapped in the circuit.
    - Always ensure the system is sealed and that the manifold gauge is properly connected before proceeding with diagnostics.

6.3 Analysis of Operating Parameters with a Thermal Camera

The use of a thermal camera is one of the most effective techniques for analyzing the operation of an HVAC system without having to disassemble components or interrupt the operational cycle. Thermography allows for the detection of temperature variations, identifying inefficiencies in the system, abnormal heat accumulation, insulation leaks, and malfunctions in electrical components.

In the United States, thermal camera technology is widely used not only by HVAC technicians but also by building inspectors and energy efficiency specialists. Thanks to its ability to visualize temperature differences through infrared images, a thermal camera helps quickly identify problems that might not be visible to the naked eye.

How a Thermal Camera Works in HVAC Systems

A thermal camera captures infrared radiation emitted by objects and converts it into a thermal image, where each color represents a different surface temperature. Warmer components appear in shades of red, orange, and yellow, while cooler ones are displayed in blue and purple.

For an HVAC system, a thermal camera allows for the analysis of heat distribution in ducts, air handling units, evaporator, and condenser coils, immediately identifying anomalies that could indicate a problem.

An overheating compressor, a frozen evaporator, or insulation leaks in piping can be detected with a simple thermographic scan without requiring invasive disassembly.

Detecting Refrigerant Leaks and Coil Issues

One of the most common problems in HVAC systems is refrigerant leaks, which can reduce system efficiency and damage the compressor. A refrigerant leak causes abnormal temperatures in the piping and evaporator coils, which can be easily detected with a thermal camera.

If an evaporator coil appears colder than normal or shows an uneven temperature distribution, there may be a restriction in refrigerant flow or an issue with the thermostatic expansion valve.

In the condenser, an anomaly in heat distribution may indicate insufficient ventilation, dirt accumulation on the heat exchange fins, or a pressure problem in the refrigerant circuit.

Analysis of Insulation Leaks and Heat Distribution

An efficient HVAC system must evenly distribute heat or cooling throughout the building. If some rooms are noticeably colder or warmer than others, there may be an insulation issue in the air ducts or the building structure.

Using a thermal camera, it is possible to analyze thermal dispersion in walls, ceilings, and floors, identifying areas where heat is escaping or accumulating. This helps pinpoint critical spots that require improved insulation or duct adjustments.

If air vents show significant temperature differences, the problem could be an air leak in the ducts, an internal obstruction, or an improper system balance.

Diagnosing Electrical Components and Preventing Failures

Beyond mechanical component analysis, a thermal camera is highly useful for detecting abnormal overheating in electrical panels, motors, and compressors.

A faulty wiring connection or a loose connector can generate electrical resistance, leading to increased temperature and potential short circuits or sudden failures.

If a motor or capacitor appears significantly hotter than usual compared to other components, this may indicate overload, low voltage, or lubrication issues in the bearings.

In industrial and commercial HVAC systems, thermography is used for predictive maintenance, helping prevent failures before they become critical.

How to Use a Thermal Camera for HVAC Analysis

To obtain reliable data, the thermal camera must be correctly set up and used under appropriate environmental conditions.

The operator should avoid thermal reflections from metallic surfaces that can alter readings and adjust the instrument's emissivity based on the material being analyzed.

Before conducting a scan, the HVAC system should be running for at least 15-30 minutes to allow thermal stabilization of the components. Once the measurement is taken, the collected data should be compared with reference values to determine if the system is operating within proper parameters.

Using thermography in combination with other diagnostic techniques, such as refrigerant pressure measurement or electrical current analysis, provides a comprehensive assessment of the system's performance.

6.4 Duct Leakage Testing and How to Detect Air Leaks

Air leaks in HVAC ducts are one of the main factors that reduce system efficiency and increase energy costs. In the United States, it is estimated that up to 30% of conditioned air is lost due to leaking ducts or faulty joints. This not only compromises indoor comfort but can also place excessive strain on HVAC units, shortening their operational lifespan.

Identifying and repairing air leaks is essential for improving system efficiency, reducing energy consumption, and ensuring uniform air distribution throughout a home or commercial building. To conduct an accurate inspection, manual methods, electronic tools, and advanced testing techniques using thermographic or differential pressure technology can be used.

Signs of Leaks in HVAC Ducts

Duct leaks can manifest in several ways. If some rooms receive less air than others, if dust accumulates more quickly than usual, or if energy bills are unusually high, it is likely that air is escaping from the ducts.

An obvious sign of a leak is inconsistent temperatures across different rooms. If conditioned air does not reach all areas of the building properly, it means that part of the airflow is being lost along the way.

A hissing sound or air movement around the ducts may also indicate a problem. In some cases, leaks occur at the joints between duct segments or where the ducts pass through walls and ceilings.

Manual Inspection and Smoke or Incense Test

The simplest method to check duct integrity is a visual and manual inspection. Running a hand along the joints and bends of the ducts while the system is running can help detect any escaping air.

A basic home method involves using an incense stick or a smoke generator near duct joints. If the smoke is pushed away or pulled into a specific spot, it is likely that there is a leak.

These methods can be useful for detecting obvious leaks, but they are not precise enough to identify small air losses that can still have a significant impact on HVAC system performance.

Using Thermography to Identify Duct Leaks

A thermal camera allows for the visualization of temperature differences along the ducts, revealing areas where conditioned air escapes before reaching the vents.

A properly sealed duct will show a uniform temperature distribution, while a leaking duct will have hot or cold spots compared to the rest of the system.

Thermography is particularly effective for detecting leaks in ducts hidden within ceilings or walls, where a manual inspection is not possible.

Advanced Testing: Pressure Testing with the Blower Door Test

For a more detailed duct integrity assessment, a Blower Door Test can be performed. This method is used by HVAC technicians to measure air leakage in buildings.

The test involves using a specialized fan installed at the main entrance to create a pressure difference between the interior and exterior of the building. This pressure variation highlights duct leaks and quantifies air loss.

In addition to the Blower Door Test, some technicians use a Duct Leakage Tester, a device that connects directly to HVAC ducts and precisely measures leakage levels.

Sealing Leaks and Optimizing the HVAC System

Once leaks have been identified, faulty joints can be sealed using HVAC-certified duct sealing tape or specialized duct mastic.

It is important to avoid using regular adhesive tape or generic sealants, as they may degrade over time due to temperature fluctuations and humidity.

For a durable seal, it is recommended to apply a layer of mastic over the connections and then cover them with aluminum sealing tape. This method ensures a tight seal and prevents future leaks.

In modern HVAC systems, the use of electronic sensors and real-time monitoring systems helps continuously track duct integrity and detect anomalies before they become serious issues.

6.5 Software and Apps for HVAC System Management and Monitoring

In recent years, the HVAC sector has evolved rapidly thanks to the integration of advanced software and mobile applications that allow real-time monitoring, management, and optimization of systems. These tools provide significant benefits for both HVAC professionals and residential and commercial building owners, enabling improved energy efficiency, lower operating costs, and proactive maintenance.

In the United States, digital solutions for HVAC system monitoring have become increasingly popular due to government incentives for reducing energy consumption and the growing demand for home automation. With dedicated apps and advanced diagnostic software, users can perform remote checks, analyze system performance, and receive real-time notifications in case of anomalies.

Monitoring Software for Commercial and Residential HVAC Systems

Companies and large building owners use HVAC management software to oversee system operations across multiple locations and optimize energy use. These systems are generally integrated with Building

Management Systems (BMS) or Building Automation Systems (BAS), advanced platforms that centralize the control of heating, ventilation, and air conditioning.

Through an intuitive graphical interface, technicians can monitor key parameters such as temperature, humidity levels, energy consumption, and the status of major components. Some software solutions use artificial intelligence algorithms to predict malfunctions and recommend preventive maintenance actions.

The most advanced solutions also include IoT (Internet of Things) sensors, which collect real-time data from each HVAC unit and transmit it to a centralized cloud, allowing the system to adjust its operation based on environmental conditions.

Mobile Applications for Remote Control and Diagnostics

Mobile apps are transforming the way HVAC system owners manage their units. Many air conditioning manufacturers offer dedicated apps that allow users to adjust temperature settings, schedule operating times, and receive system alerts directly on their smartphones.

The most advanced apps provide remote diagnostics, enabling users to check system status and receive troubleshooting recommendations before contacting a technician. Some smart thermostats, such as Nest, Ecobee, and Honeywell, use artificial intelligence to learn user habits and automatically adjust temperature settings based on occupancy and external weather conditions.

These applications are particularly useful for vacation homes or commercial properties, as they allow remote climate control and reduce energy consumption when spaces are unoccupied.

Digital Tools for HVAC Diagnostics and Maintenance

HVAC professionals can leverage advanced diagnostic software that simplifies troubleshooting and maintenance management. Some digital tools connect directly to HVAC systems via Bluetooth or Wi-Fi, gathering real-time data on pressure levels, temperatures, and fan speeds.

There are also tools that allow technicians to generate digital maintenance reports, automatically logging all performed operations and alerting them when service is due. Some software solutions incorporate augmented reality, overlaying virtual instructions directly onto system components to guide technicians through repairs.

Using these tools reduces response times, improves diagnostic accuracy, and optimizes technical resource management.

Integration with Smart Home Systems and Energy Savings

The integration of HVAC systems with smart home technology is becoming increasingly common in the United States, especially with the rise of home automation. Modern HVAC units can connect to voice assistants like Amazon Alexa, Google Assistant, and Apple HomeKit, allowing users to adjust temperatures with simple voice commands.

Smart thermostats play a crucial role in this integration, as they analyze weather data, occupancy patterns, and energy consumption in real-time to optimize system usage. Some advanced models send notifications when a filter replacement is needed or when they detect abnormal system behavior.

HVAC systems connected to home automation networks can also interact with solar panels and home battery storage, adjusting temperature settings to maximize efficiency and reduce energy costs.

Benefits of Automation in Preventive Maintenance

One of the biggest advantages of using software and apps for HVAC system monitoring is the ability to perform preventive maintenance based on real-time system data.

Through machine learning algorithms, some software can identify energy consumption trends and predict failures before they occur, allowing users to schedule service before issues become critical.

A digitally managed HVAC system can also automatically adapt to changing climate conditions, adjusting airflow and heating or cooling power based on outdoor humidity and temperature.

With these tools, users can reduce operational costs, enhance indoor comfort, and extend system lifespan. The adoption of these technologies is continuously growing, with an increasing number of commercial and residential buildings choosing digital solutions to optimize energy efficiency.

BOOK 7
Common Repairs: Step-by-Step Guide

7.1 Replacing a Faulty Capacitor

The capacitor is an essential component in HVAC systems, responsible for starting and maintaining the operation of the compressor and fans. When it fails, the unit may struggle to start, operate inefficiently, or, in the worst cases, fail to turn on at all.

In the United States, many HVAC failures are caused by faulty capacitors, especially in regions with hot summers where cooling systems undergo heavy use. Fortunately, replacing a capacitor is a relatively simple procedure that can be performed with basic tools and proper safety precautions.

Symptoms of a Faulty Capacitor

Before replacing a capacitor, it is necessary to confirm that it is the actual cause of the problem. Common signs of capacitor failure include:

- The HVAC unit does not start or takes longer than usual to power on.
- A humming sound comes from the outdoor unit, but the compressor or fans do not start.
- The system turns on but shuts down unexpectedly after a few minutes.
- The compressor or fan runs slower than usual, reducing cooling efficiency.
- The capacitor appears visibly swollen or damaged.

If one or more of these symptoms are present, the capacitor should be tested to confirm failure before proceeding with the replacement.

Testing the Capacitor with a Multimeter

Before removing the capacitor, it should be tested to ensure it is no longer functional. The best way to do this is by using a digital multimeter with a capacitance measurement function (μF).

- Turn off the system and disconnect the power supply to avoid the risk of electric shock.

- Discharge the capacitor by using an insulated-handled screwdriver to connect both terminals and dissipate any residual energy.
- Set the multimeter to capacitance mode (µF) and connect the probes to the capacitor terminals.
- Read the value on the display: if the reading is lower than the rating indicated on the capacitor's label, the component is faulty and needs replacement.
- If the multimeter does not show any reading or indicates an open circuit, the capacitor is completely defective and can no longer store a charge.

Removing the Faulty Capacitor

Once confirmed that the capacitor is defective, it can be removed.

- Locate the capacitor inside the HVAC unit. It is usually found in the control panel of the outdoor unit.
- Note the position and wiring connections before disconnecting them, using labels or taking a photo for reference.
- Disconnect the terminals using pliers, being careful not to damage the wires.
- Remove the capacitor by unscrewing the mounting brackets.

Installing the New Capacitor

Before installing the new capacitor, it is crucial to verify that it has the same specifications as the old one. The capacitance in microfarads (µF) must be identical, and the rated voltage must be equal to or higher than that of the original capacitor.

- Place the new capacitor into the mounting bracket and secure it with screws.
- Connect the wires to the correct terminals, using the wiring diagram or the previously taken photo as a reference.
- Ensure the terminals are tightly secured to prevent poor connections or electrical arcing.

System Test and Final Verification

After installing the new capacitor, it is important to test the system to ensure it is working correctly.

- Restore the power supply and turn on the HVAC unit.
- Listen for the startup sound: if the capacitor is functioning properly, the compressor and fans should start without difficulty.
- Check the temperature of the air coming from the vents to confirm that the system is cooling or heating effectively.
- Monitor the system for a few minutes to ensure there are no sudden shutdowns or unusual noises.

If the unit starts without issues and operates efficiently, the replacement was successful. However, if the problem persists, other components such as the compressor or the start relay may be faulty.

7.2 Repairing a Stuck Fan

The fan in an HVAC system plays a crucial role in ensuring proper airflow through the unit. If the fan gets stuck or stops working, the system will not be able to dissipate heat or distribute conditioned air efficiently, leading to overheating and potential damage to the compressor.

In the United States, where HVAC systems operate under extreme climates, a stuck fan can quickly become a critical issue, resulting in complete system failure. Identifying the cause of the blockage and addressing it promptly prevents more severe damage and costly repairs.

Signs of a Stuck or Malfunctioning Fan

A fan issue can present itself in various ways. If the HVAC system is running but no airflow is coming from the vents, or if a humming noise is heard from the outdoor unit without the fan blades spinning, a blockage is likely.

In some cases, the fan motor may attempt to start, but the blades spin slowly or do not move at all. If a burning smell or overheating odor is detected near the unit, the problem may be related to the motor or internal bearings.

Diagnosing the Cause of the Blockage

Before proceeding with repairs, it is essential to determine why the fan is stuck. Some of the most common causes include:

- Accumulation of dirt and debris: Leaves, dust, and other particles can enter the outdoor unit and obstruct the movement of the blades.
- Worn-out bearings: Over time, the fan motor bearings can degrade, increasing resistance to movement.
- Faulty capacitor: If the capacitor that powers the fan motor is defective, the motor may not receive enough energy to start.
- Electrical issues: Damaged wiring, a blown fuse, or a faulty relay can prevent the motor from receiving power.

To determine whether the issue is mechanical or electrical, try manually rotating the fan blades while the system is off. If the blades spin freely, the problem is likely electrical. If they feel stiff or stuck, the issue is mechanical.

Manually Unlocking and Cleaning the Fan

If debris is blocking the fan, it must be removed, and the unit should be cleaned to prevent future issues.

- Turn off the HVAC unit's power supply to ensure safety.
- Remove the outdoor unit cover or access panel to reach the fan.
- Inspect the fan for any debris or dirt buildup.
- Use a soft brush and compressed air to remove dust and residue.
- Manually rotate the blades to ensure they move without resistance.

After cleaning, turn the system back on to check if the fan resumes normal operation.

Replacing the Fan Motor Bearings

If the fan blades remain difficult to move even after cleaning, the bearings may be worn out and require lubrication or replacement.

- Remove the fan motor by unscrewing the mounting brackets and disconnecting the power cables.
- Apply electric motor lubricant to the bearings, rotating the shaft to distribute the lubricant evenly.
- If the bearings are too worn out, replace the fan motor with a compatible model.
- Reinstall the motor and reconnect the electrical cables.

Once the replacement is complete, test the system to ensure the fan spins smoothly.

Testing the Fan Motor and Start Capacitor

If the fan still does not function after cleaning and lubrication, the problem could be electrical. To test the motor, use a multimeter:

- Set the multimeter to resistance mode (Ω).
- Disconnect the motor from the electrical circuit.
- Connect the multimeter probes to the motor terminals and check if the resistance value falls within the manufacturer's specifications.
- If the resistance is too high or the circuit is open, the motor is faulty and must be replaced.

Similarly, the start capacitor must be tested to ensure it provides the necessary energy to the motor. If the capacitor is defective, it must be replaced with one that has the same capacitance and voltage rating.

7.3 Replacing a Faulty Relay

The relay is a crucial electrical component in an HVAC system. It acts as an automatic switch, allowing current to flow to various electrical components, such as the compressor and fans, when requested by the thermostat. A faulty relay can prevent the system from starting, cause sudden shutdowns, or lead to irregular operation of the compressor and fan.

In the United States, where HVAC systems are used daily for both heating and cooling, promptly replacing a defective relay is essential to avoid disruptions in home comfort. Fortunately, this is a relatively simple task that can be performed with a few tools, as long as the necessary safety precautions are followed.

Symptoms of a Faulty Relay

When a relay malfunctions, the symptoms can vary depending on the type of failure. If the relay is completely burned out, the HVAC system may not turn on at all. If it is defective but still partially functional, the compressor or fan may start and stop irregularly.

Another common sign is a buzzing or repeated clicking sound coming from the electrical panel of the HVAC unit. If the relay attempts to activate but fails to maintain electrical contact, these unusual noises may be heard.

In more severe cases, a damaged relay can get stuck in the closed position, causing the compressor or fan to run continuously, even when the thermostat is not calling for cooling or heating. This can lead to overheating and increased energy consumption.

Testing the Relay with a Multimeter

Before replacing the relay, it is essential to verify that it is indeed the faulty component. The most effective way to do this is by using a digital multimeter to test electrical continuity.

- Turn off the HVAC system and disconnect it from the power supply to avoid the risk of electric shock.
- Access the electrical panel of the outdoor unit and locate the relay. It is usually found near the contactor or control panel.
- Set the multimeter to continuity mode (symbol of a buzzer or Ω).
- Connect the multimeter probes to the relay terminals and check whether the circuit is open or closed.
- Activate the thermostat to see if the relay closes when the system is turned on.
- If the multimeter does not detect continuity when the relay should be active, it means the component is faulty and must be replaced.

Removing the Faulty Relay

Once it has been confirmed that the relay is defective, it can be removed.

- Locate the relay and note the wire connections. Taking a picture can be helpful for reference during installation.
- Disconnect the electrical terminals using pliers, being careful not to damage the wires.
- Remove the relay from its housing by unscrewing the mounting screws or unlocking it from the support.
- Inspect the wiring to ensure there are no signs of burns or damage. If necessary, replace any worn connectors as well.

Installing the New Relay

Before installing the new relay, it is essential to verify that it has the same voltage and amperage specifications as the original component. A relay with incorrect characteristics may not function properly or could damage other electrical components in the system.

- Insert the new relay into the mounting support and secure it firmly.
- Connect the wires to the corresponding terminals, following the original wiring diagram.
- Ensure that all connections are tight and free of oxidation.

- If the relay is a plug-in type, make sure it is properly inserted into the socket without forcing the contacts.

System Test and Final Verification

After installation, the HVAC system must be tested to confirm that the new relay is functioning correctly.

- Restore power to the system and turn on the thermostat.
- Listen to the relay: it should produce a single click when the system starts, without repeated activation attempts.
- Check the operation of the compressor and fans. If everything starts up without issues, the replacement was successful.
- Monitor the system for a few minutes to ensure there are no sudden shutdowns or abnormal operations.

If the problem persists after replacing the relay, it may be necessary to check the contactor or control panel to identify any other electrical faults in the circuit.

7.4 How to Repair an HVAC System with a Refrigerant Leak

A refrigerant leak is one of the most critical issues that can affect an HVAC system. Refrigerant is the fluid responsible for transferring heat between the indoor and outdoor units. If the level drops below the optimal amount, the system will struggle to cool or heat efficiently, leading to increased energy consumption and, in severe cases, irreversible damage to the compressor.

In the United States, refrigerant handling is regulated by the Environmental Protection Agency (EPA), and only certified technicians can perform recharges or repairs involving the release of gas into the environment. However, identifying a leak and taking necessary precautions for repairs can be done by those with a good understanding of HVAC systems.

Signs of a Refrigerant Leak

An HVAC system losing refrigerant may exhibit several noticeable symptoms. If the air coming from the vents is not sufficiently cold, the issue may be due to low refrigerant levels.

Another common sign is the formation of ice on the evaporator coils or refrigerant lines. This happens because an insufficient amount of refrigerant causes excessive temperature drops in the system, leading to condensation and subsequent ice buildup.

If the outdoor unit emits a hissing or gurgling sound, this could indicate a leak. In severe cases, the compressor may overheat and frequently shut down as it struggles to compensate for the refrigerant loss.

Detecting the Refrigerant Leak

There are several methods for identifying the source of a refrigerant leak. The first step is a visual inspection of the refrigerant circuit pipes, particularly at the connections and welds, where leaks most commonly occur.

If there are no visible signs of oil or moisture around the pipes, an electronic leak detector can be used. These devices, also known as "sniffers," can detect refrigerant gas in the air and pinpoint the source of the leak.

Another effective method is the soapy water test. By applying a mixture of water and soap to joints and pipes, bubbles will form in the presence of a leak.

For leaks too small to be detected with traditional methods, a pressurized nitrogen test can be performed. This process involves evacuating the system and refilling it with high-pressure nitrogen to check for pressure drops, indicating a leak.

Repairing the Leak

Once the leak has been identified, the next step is to perform the repair. If the issue is caused by a loose fitting, tightening the connection with a torque wrench may resolve the problem.

If the leak comes from a faulty weld, a gas torch must be used to re-solder the leak point. It is important to thoroughly clean the area before soldering to ensure optimal material adhesion.

For leaks in corroded or damaged pipes, the defective section must be removed and replaced with a new copper tube, which should be brazed into place and retested for leaks.

If the leak is very small and does not justify a full repair, a refrigerant sealant can be used. These products are introduced into the system and automatically seal micro-leaks without affecting system performance. However, this is recommended only as a temporary solution.

Recharging the Refrigerant and Testing the System

After the repair, the system must be evacuated to remove moisture and contaminants. This process is done using a vacuum pump, which creates negative pressure inside the system to ensure no air residues remain before recharging.

Next, refrigerant can be added through a service valve, following the manufacturer's specifications to ensure the correct level is restored. It is crucial to use the type of refrigerant specified on the HVAC unit label to avoid chemical incompatibilities that could damage the compressor.

After recharging, the system should be tested to ensure it is functioning properly. This involves checking the pressure with an HVAC gauge and measuring the air temperature coming from the vents.

Preventing Future Leaks

Refrigerant leaks can be prevented with regular maintenance. Inspecting the pipes and fittings at least once a year helps detect early signs of corrosion or wear before significant leaks occur.

Another key precaution is to avoid overloading the system by setting the thermostat to maintain an efficient temperature without excessive fluctuations. Frequent temperature changes and continuous on-off cycles can accelerate wear on welds and joints.

Additionally, installing a dehumidifier filter helps prevent moisture buildup inside the refrigerant circuit, reducing the risk of internal corrosion.

7.5 Repairing a Faulty Thermostat

The thermostat is the control center of an HVAC system. It regulates indoor temperature by turning heating or cooling on and off based on user settings. When the thermostat malfunctions, the entire system may stop working properly, causing temperature fluctuations, high energy consumption, or the inability of the HVAC unit to turn on.

In the United States, where HVAC systems are essential for handling both hot summers and cold winters, a faulty thermostat can quickly become a serious issue for home comfort. Identifying the fault and repairing it promptly is crucial for maintaining efficient climate control.

Signs of a Faulty Thermostat

If the thermostat is not working correctly, several issues may arise. A common sign is the HVAC system failing to turn on despite the thermostat being set to the correct temperature. If the unit remains inactive even when the temperature is adjusted higher or lower, the issue could be an electrical failure or a malfunctioning internal sensor.

Another symptom is continuous system operation without shutting off, even after reaching the set temperature. This happens when the thermostat fails to accurately detect room temperature, causing the HVAC system to work longer than necessary and increasing energy consumption.

If the thermostat display is off or showing incorrect information, there may be a power supply issue. In wired thermostats, this could be due to a faulty electrical connection, while in battery-operated models, the most likely cause is a dead or corroded battery.

Diagnosing a Thermostat Malfunction

Before replacing the thermostat, it is important to check that the issue is not caused by an incorrect setting or a power interruption. The first step is to ensure that the thermostat is turned on and set to the correct mode (heating or cooling).

If the thermostat is battery-powered, replace the batteries with a new set and check if the issue is resolved. If it is a wired thermostat, verify that the wires are properly connected and that no fuses are blown in the HVAC system's electrical panel.

To determine if the thermostat is sending the correct signal to the HVAC unit, a multimeter can be used to test the voltage at the terminals. If the thermostat is not providing the correct voltage, it means the component is faulty and needs to be replaced.

Repairing and Resetting the Thermostat

If the thermostat appears to be on but is not communicating with the HVAC system, a reset can be attempted to restore factory settings. The procedure varies by model but usually involves pressing a reset button or removing the batteries for a few minutes before reinstalling them.

If the issue is due to dust or dirt buildup inside the thermostat, the cover can be removed, and the sensor can be gently cleaned with a soft-bristle brush. Dust can interfere with the sensor's ability to accurately detect temperature, causing incorrect readings.

For older mechanical thermostats, the problem may be a faulty switch or a damaged coil. If the thermostat does not respond after a reset and cleaning, replacing it with a newer programmable model may be a better option.

Replacing a Faulty Thermostat

If the thermostat is beyond repair, it needs to be replaced with a compatible model. When choosing a new thermostat, it is essential to ensure that it matches the type of HVAC system installed, whether it is a single-stage, two-stage, or multi-zone system.

To install a new thermostat:

- Turn off the power to the HVAC system to prevent short circuits or electric shocks.
- Remove the old thermostat by unscrewing the front panel and disconnecting the wires from the terminals. Taking a picture of the wiring setup is recommended to simplify installation.
- Mount the new thermostat base, ensuring it is level and securely attached to the wall.
- Connect the wires to the new thermostat, following the wiring diagram provided by the manufacturer.
- Turn the HVAC system back on and test the new thermostat to confirm that it communicates properly with the unit.

If installing a smart thermostat, it may be necessary to connect it to Wi-Fi and configure it via a mobile app to access advanced automation and remote control features.

Preventing Future Malfunctions

To avoid future thermostat issues, periodic maintenance is recommended. Regularly cleaning the temperature sensor and avoiding thermostat placement in drafty areas, near windows, or heat sources can help ensure accurate readings.

If using a battery-powered thermostat, replace the batteries at least once a year to maintain continuous operation. In wired systems, periodically checking electrical connections can prevent contact issues or oxidation.

Smart thermostats offer diagnostic tools that allow users to monitor HVAC system performance and receive notifications in case of problems. Utilizing these technologies helps prevent unexpected failures and keeps the system running efficiently.

BOOK 8
Working in the HVAC Industry: Career and Certifications

8.1 Skills Required to Become an HVAC Technician

The HVAC industry is one of the most in-demand fields in the United States, with a constant need for qualified technicians to install, maintain, and repair heating, ventilation, and air conditioning systems. To succeed in this field, it is essential to develop a combination of technical skills, theoretical knowledge, and practical abilities, along with a strong aptitude for problem-solving.

Working as an HVAC technician requires specialized training, which can be obtained through professional courses, apprenticeship programs, and official certifications. However, beyond technical preparation, a good technician must also possess communication, organizational, and attention-to-detail skills to diagnose and resolve problems efficiently.

Understanding HVAC System Operations

One of the fundamental skills of an HVAC technician is the ability to understand how different air conditioning systems function, from residential units to large commercial installations. It is essential to know the thermodynamic cycle, the role of the compressor, condenser, evaporator, and expansion valves, as well as the principles of thermodynamics and heat transfer.

In addition to theoretical knowledge, a technician must be able to interpret electrical schematics and flow diagrams, understanding how different components interact. This ability is crucial for quickly identifying faults and inefficiencies in HVAC systems.

Diagnostic and Problem-Solving Skills

One of the primary responsibilities of an HVAC technician is diagnosing issues and finding effective solutions. Every service call requires a thorough analysis to determine the cause of the malfunction, whether it is a refrigerant leak, an electrical failure, or a mechanical issue.

To do this, a technician must be skilled in using advanced diagnostic tools such as multimeters, refrigerant pressure gauges, thermal cameras, and leak detectors. Knowing how to interpret the data from these tools is essential for quickly identifying problems and providing the appropriate solutions.

Practical Skills in Installation and Maintenance

A good HVAC technician must have hands-on experience in installing new systems and performing preventive maintenance on existing ones. This includes mounting indoor and outdoor units, connecting refrigerant lines, installing thermostats, and configuring ventilation systems.

Maintenance is equally important, as it helps prevent breakdowns and extends the lifespan of the system. Key tasks include cleaning coils, replacing air filters, checking refrigerant pressure, and inspecting electrical contacts.

To perform these operations effectively, a technician must be proficient in using specialized tools such as brazing torches, vacuum pumps, and airflow balancing instruments.

Knowledge of Electrical and Electronic Systems

Modern HVAC systems incorporate advanced electronic controls, temperature sensors, and smart thermostats. A technician must be able to work with electrical circuits, test component continuity, identify short circuits, and replace relays, capacitors, and electric motors.

A solid understanding of electrical schematics helps diagnose and repair faults in the connections between the thermostat, contactor, and power supply units. Knowing how to read and interpret an HVAC electrical diagram is essential for quickly identifying circuit issues.

Safety and Regulatory Compliance Skills

Working in the HVAC industry involves risks related to electricity, refrigerant gases, and work tools. Therefore, a technician must be familiar with and adhere to OSHA (Occupational Safety and Health Administration) safety regulations, as well as EPA regulations on the handling and usage of refrigerants.

Wearing PPE (Personal Protective Equipment) such as insulated gloves, safety glasses, and masks to protect against welding fumes is essential to prevent injuries. Additionally, a technician must know how to properly handle refrigerant gases without releasing them into the environment, using refrigerant recovery pumps and complying with environmental regulations.

8.2 Certifications and Licenses Required to Work in the HVAC Industry

Becoming an HVAC technician in the United States is not just about acquiring technical and practical skills. To legally operate in the field, it is essential to obtain the certifications and licenses required by state and federal regulations. These credentials demonstrate that a technician has the necessary knowledge to work with refrigerants, electrical systems, and HVAC equipment safely and in compliance with environmental regulations.

In the U.S., regulations vary by state, but there are several nationally recognized certifications that are essential for any HVAC professional. Obtaining these qualifications not only ensures access to job opportunities but also allows for career advancement and the possibility of starting an independent business.

EPA Section 608 – Refrigerant Handling Certification

One of the most important certifications for HVAC technicians is the EPA Section 608, required by the Environmental Protection Agency (EPA) for anyone working with systems that contain regulated refrigerants. This certification is mandatory for handling, recovering, and disposing of refrigerant gases and is divided into four categories:

- Type I: For technicians working on small appliances such as refrigerators and portable air conditioners.
- Type II: For those operating on high-pressure HVAC systems, including residential and commercial units.
- Type III: For technicians specializing in low-pressure systems, such as industrial refrigeration units.
- Universal: Covers all the previous certifications in a single exam.

To obtain the EPA 608 certification, candidates must pass a written exam covering refrigerant theory, environmental regulations, and safe handling practices for fluorinated gases.

NATE – North American Technician Excellence

Another crucial certification for HVAC technicians is NATE (North American Technician Excellence), recognized nationwide as a standard of excellence in the industry. Unlike the mandatory EPA 608 certification, NATE is voluntary, but obtaining it increases the chances of being hired by top companies and earning a higher salary.

The NATE test covers a wide range of topics, including:

- HVAC system diagnosis and repair
- Installation procedures
- Workplace safety
- Energy efficiency

There are multiple levels of NATE certification, such as the Certified HVAC Technician (CHT) for entry-level technicians and the Senior Level Efficiency Analyst for more experienced professionals.

State-Specific Certifications and Local Licenses

In addition to national certifications, each state has specific requirements for obtaining an HVAC license. Some states require technicians to pass a state exam before they can work independently or start their own business.

For example:

- Texas, Florida, and California require an HVAC state license for installing or repairing systems.
- Colorado allows technicians to work under the supervision of a certified professional without needing an individual license.

Before starting work, it is essential to check the specific requirements in the state where you plan to operate. This information can be obtained from the Department of Labor or professional licensing regulatory offices.

OSHA – Workplace Safety Certification

HVAC systems involve risks related to electricity, high temperatures, and chemical handling. For this reason, many employers require HVAC technicians to obtain an OSHA (Occupational Safety and Health Administration) certification, which provides workplace safety training.

The most common OSHA courses for HVAC professionals include:

- OSHA 10-Hour Training: For entry-level technicians who need basic safety training.
- OSHA 30-Hour Training: For more experienced professionals who manage worksites and teams.

These certifications help prevent workplace accidents and ensure compliance with federal safety regulations.

Other Specialized Certifications

Beyond the mandatory certifications, there are advanced credentials that allow HVAC technicians to specialize in specific areas. Some examples include:

- HVAC Excellence Certification: An advanced certification program for experienced technicians.
- R-410A Certification: Required for working with R-410A refrigerant, which is increasingly used in modern HVAC systems.
- Building Performance Institute (BPI) Certification: Ideal for those specializing in energy efficiency and building performance improvement.

Obtaining these additional certifications can increase professional credibility and improve career opportunities.

8.3 How to Start a Career or Business in HVAC

The HVAC industry offers numerous career opportunities, both for those who want to work as specialized technicians and for those who aim to start their own business. With the growing demand for energy-efficient systems and the continuous need for maintenance and repairs, pursuing a career in this field can ensure financial stability and growth opportunities.

To successfully enter the HVAC field, it is essential to follow a structured path, including training, certifications, practical experience, and, in the case of starting a business, a solid business plan. Each step must be carefully planned to avoid common mistakes that could slow down professional success.

Training and Specialization

The first step in entering the HVAC industry is acquiring a solid technical foundation. This can be achieved through trade schools, technical training programs, or apprenticeships with HVAC companies. In the

United States, many schools offer specialized HVAC courses covering topics such as thermodynamics, installation and maintenance of air conditioning systems, and environmental regulations.

Beyond basic training, specializing in specific areas such as commercial refrigeration, industrial ventilation systems, or energy efficiency technologies can increase competitiveness in the job market and offer higher earning potential.

A great way to gain experience is by working as an apprentice at an HVAC company, assisting experienced technicians to learn directly in the field. This hands-on training allows for the development of practical skills and a deeper understanding of HVAC systems in real-world applications.

Obtaining Required Certifications

In the United States, obtaining certain certifications is mandatory to legally work as an HVAC technician. The most important is the EPA Section 608 certification, required for handling refrigerants in compliance with environmental regulations. Getting this certification is an essential step for anyone working on air conditioning systems.

Other certifications, such as NATE (North American Technician Excellence), can enhance career opportunities, making technicians more qualified in the eyes of employers and clients. Depending on the state, an HVAC state license may also be required, which could involve passing a written and practical exam.

Working as an HVAC Technician: Opportunities and Career Paths

After completing training and obtaining certifications, the next step is finding a job in the industry. Many HVAC technicians start by working for specialized companies that focus on system installation and maintenance, gaining experience and practical skills.

There are various career opportunities, including:

- Working in residential service companies
- Specializing in large commercial or industrial HVAC systems
- Focusing on refrigeration system maintenance
- Installing HVAC systems in newly constructed buildings

With experience, it is possible to advance in the career and take on higher responsibility roles, such as HVAC supervisor or project manager. Some professionals choose to become HVAC inspectors, ensuring system compliance with safety and energy efficiency regulations.

Starting an HVAC Business

Many HVAC technicians, after gaining experience, decide to start their own business. Running an HVAC company can be highly profitable but requires careful planning and solid financial management.

The first step is obtaining a business license and proper insurance coverage, which are necessary to operate legally and protect the business from potential liabilities. Choosing a specialization, such as residential installations, industrial maintenance, or energy efficiency solutions, can help differentiate the business from competitors.

Building a strong client network is crucial for success. Using marketing strategies such as online advertising, word-of-mouth referrals, and a professional website can help promote services and attract new customers.

Another important aspect is financial management. Accurately calculating operational costs, setting competitive rates, and planning investments in equipment and tools will ensure the long-term sustainability of the business.

Staying Updated and Growing in the Industry

The HVAC industry is constantly evolving, with new technologies and regulations changing frequently. To remain competitive, continuous training and staying informed about industry innovations are essential.

Attending update courses, obtaining additional certifications, and keeping up with trends like HVAC integration with smart home systems and eco-friendly refrigerants can provide competitive advantages and new growth opportunities.

Another crucial element is developing strong customer service skills. Being reliable, transparent, and communicative with customers helps build a solid reputation, which is essential for securing referrals and retaining clients.

8.4 Mistakes to Avoid as an HVAC Technician

Becoming a successful HVAC technician is not just about acquiring technical skills and obtaining certifications; it also requires avoiding common mistakes that could compromise work efficiency, customer satisfaction, and even personal safety.

In the HVAC industry, small errors can have significant consequences, ranging from system malfunctions to costly equipment damage and even health and safety risks. Therefore, it is essential to develop good work habits early in one's career and pay close attention to every detail during installation, maintenance, and repair of HVAC systems.

Ignoring a Complete Diagnosis of the Problem

One of the most common mistakes an HVAC technician can make is focusing only on the symptom of the problem without performing a complete diagnosis. For example, if a system is not cooling properly, immediately replacing the refrigerant without checking for potential leaks may lead to a temporary solution that does not address the root cause of the malfunction.

An accurate diagnosis requires examining the entire system, testing key components, and checking for anomalies in the electrical circuit or refrigerant pressure levels. Using diagnostic tools such as multimeters, thermal cameras, and leak detectors can help identify the true cause of the issue, preventing incorrect interventions and unnecessary repairs.

Failing to Follow Safety Procedures

The HVAC industry involves significant risks, including exposure to refrigerant gases, contact with high-voltage electrical systems, and handling heavy tools. Ignoring safety procedures can lead to severe accidents, ranging from chemical burns to electrical short circuits.

Always wearing Personal Protective Equipment (PPE) such as insulated gloves, protective goggles, and masks to prevent inhaling harmful gases is a fundamental rule. Additionally, before working on an HVAC unit, it is crucial to disconnect the power supply and check for the absence of voltage using a multimeter.

Proper refrigerant recovery and disposal are also critical for personal safety and compliance with EPA regulations. Avoid improper practices, such as releasing refrigerant into the atmosphere, which is not only illegal but can also cause environmental damage and health problems.

Incorrect Installation of Key Components

Poor installation can compromise the efficiency of an HVAC system and lead to premature failures. A common mistake is improper thermostat placement—if installed in direct sunlight or near a heat source, it may detect incorrect temperatures, causing the system to operate inefficiently.

Errors in mounting indoor and outdoor units can also reduce performance. An outdoor unit installed in a poorly ventilated space may overheat and work under strain, reducing energy efficiency and accelerating compressor wear.

Incorrect electrical wiring connections are another frequent issue. A wiring mistake can lead to short circuits or malfunctions that require costly repairs. Always double-check electrical connections before turning the system on.

Underestimating the Importance of Preventive Maintenance

Many HVAC technicians focus solely on repairs, neglecting the importance of preventive maintenance. Regular system check-ups can prevent unexpected failures and extend the lifespan of the equipment.

Proper maintenance includes cleaning coils, checking air filters, verifying refrigerant levels, and inspecting electrical wiring. Ignoring these aspects can lead to reduced performance, increased energy consumption, and accelerated component wear.

Technicians who offer regular maintenance services can build customer loyalty and create a steady workflow, reducing the need for emergency repairs, which are often more costly and stressful.

Poor Communication with Customers

An HVAC technician must not only be technically skilled but also capable of communicating effectively with customers. Many mistakes arise from failing to explain repairs or from a lack of transparency regarding costs.

A customer who does not clearly understand the problem or the reason for the repair may feel uncertain and dissatisfied with the service. Explaining the issue, possible solutions, and associated costs in simple terms helps build trust and professional credibility.

Another common mistake is not providing post-service maintenance advice. Educating the customer on how to maintain their system properly—such as how often to change filters or the importance of regular check-ups—can improve the customer experience and increase the likelihood of future service calls.

8.5 Growth Opportunities in the HVAC Industry and Advanced Specializations

The HVAC industry offers a wide range of growth opportunities for technicians who want to advance their careers or specialize in niche areas. With the continuous evolution of technology and the increasing demand for more efficient and eco-friendly systems, HVAC professionals who invest in training and constant updates can access higher-paying positions, manage more complex projects, or even start their own businesses.

The HVAC industry in the United States is expanding, driven by increasingly strict energy efficiency and indoor air quality regulations, as well as the rise of smart technologies. For this reason, specializing in advanced segments of the sector can provide a more stable and competitive career.

Advanced Career Paths in the HVAC Industry

An experienced HVAC technician has several options for career advancement. After a few years of field experience, one of the most common choices is to become a supervisor or lead technician, managing teams of installers and maintenance technicians. This role requires not only advanced technical skills but also leadership and organizational abilities.

Another option is specializing in HVAC system design and engineering, working with architects and builders to develop high-efficiency climate control systems for residential and commercial buildings. This path may require additional academic training or specific certifications but offers well-paid career opportunities.

For those who prefer a more independent approach, starting their own HVAC business is an ambitious but rewarding choice. Running an HVAC installation and maintenance company allows professionals to work directly with clients, offer personalized services, and build a successful business. However, it requires skills in financial management, business planning, and marketing.

Specializing in Energy Efficiency and Low-Impact Systems

Environmental concerns and rising energy costs have led to increasing demand for HVAC technicians with expertise in energy efficiency and sustainability. Specializing in this field means acquiring knowledge about high-efficiency HVAC systems, geothermal heat pumps, eco-friendly refrigerants, and the integration of solar panels with HVAC systems.

In the United States, many companies and local governments offer incentives to improve building energy efficiency. An HVAC technician who understands LEED (Leadership in Energy and Environmental Design) certification standards and government energy-saving programs can gain a competitive edge in the market.

HVAC and Automation: Specializing in Smart Systems

The integration of smart technology into HVAC systems is one of the fastest-growing trends in the industry. Smart thermostats, remote control systems, and environmental sensors are revolutionizing how heating and cooling are managed in buildings.

Specializing in connected HVAC systems requires knowledge of electronics, computing, and communication networks. Technicians trained in this field can work with IoT (Internet of Things) systems, optimizing energy use and improving indoor comfort.

Working with smart HVAC doesn't just mean installing advanced devices; it also involves configuring software and analyzing data from sensors to maximize system efficiency. This specialization is particularly in demand in the commercial sector, where automation can lead to significant energy savings.

Predictive Maintenance and Advanced Diagnostics

The introduction of advanced diagnostic and predictive maintenance tools is transforming the HVAC industry. Technicians with experience in thermography, vibration analysis, and remote monitoring can prevent failures before they occur, reducing system downtime.

The use of thermal cameras to detect heat loss, wireless sensors to monitor air quality, and energy management software is becoming increasingly common. HVAC professionals specializing in these tools can offer premium services, attracting high-end clients and increasing the value of their work.

Career Opportunities in the Commercial and Industrial Sectors

Beyond residential systems, the commercial and industrial sectors offer growth opportunities for HVAC technicians who want to work on more complex systems. Large-scale HVAC installations used in hospitals, shopping malls, skyscrapers, and manufacturing plants require advanced skills in airflow engineering, thermal load management, and industrial automation.

Working in this sector may involve collaborating with engineers and designers, handling large-scale ventilation systems, industrial cooling systems, and customized climate control solutions for critical environments like laboratories and data centers.

BOOK 9
Smart HVAC: The Future of Climate Control

9.1 Introduction to Smart and Connected HVAC Systems

In recent years, the HVAC industry has undergone a radical transformation with the introduction of smart and connected technologies. Traditional heating, ventilation, and air conditioning systems are evolving to offer greater energy efficiency, remote control, and advanced automation. Smart HVAC systems not only improve comfort in residential and commercial environments but also reduce energy consumption and operational costs.

In the United States, where the smart home market is rapidly expanding, integrating HVAC systems with digital technologies represents a significant opportunity for industry professionals. Homes and commercial buildings are increasingly adopting advanced solutions such as smart thermostats, energy management systems, and environmental sensors to optimize real-time climate control.

What Are Smart and Connected HVAC Systems?

A smart HVAC system is an installation that uses sensors, software, and internet connectivity to automatically regulate temperature, air quality, and energy consumption based on user needs. Unlike traditional systems that require manual adjustments, a smart system can learn occupant habits, dynamically adapting to ensure maximum comfort with minimal energy waste.

Thanks to Wi-Fi or Bluetooth connectivity, these systems can be controlled and monitored remotely via smartphones, tablets, or voice assistants such as Amazon Alexa, Google Assistant, and Apple HomeKit. Users can program settings, receive notifications about system conditions, and even set automated rules to enhance efficiency.

Benefits of Smart HVAC Systems

One of the primary benefits of a smart HVAC system is energy efficiency optimization. Smart thermostats, for example, adjust the temperature based on occupancy and weather conditions, preventing energy waste when the house is empty or when outdoor temperatures allow for reduced heating or cooling loads.

Another key advantage is continuous performance monitoring. Connected systems can detect anomalies and alert users to potential issues before they become critical problems. Some advanced models can even perform self-diagnosis and send system status reports to HVAC technicians, reducing service response times and improving predictive maintenance.

Integration with air quality sensors enhances occupant well-being by automatically adjusting ventilation and filtration based on humidity levels, CO_2 concentrations, and fine particulate matter. This is particularly useful in commercial buildings and homes with sensitive individuals, such as children and the elderly.

Key Components of a Smart HVAC System

Smart HVAC systems consist of several interconnected components that work together to ensure optimal operation. The smart thermostat is the system's core, allowing temperature control and advanced scheduling. Models such as Google Nest and Ecobee SmartThermostat offer features like machine learning, geolocation, and voice commands.

Environmental sensors monitor temperature, humidity, and air quality, sending data to the HVAC system to optimize airflow and energy efficiency. Some advanced systems also use occupancy sensors, adjusting system operation based on whether a room is occupied or vacant.

Another key element is energy management software, which analyzes sensor data and suggests optimal settings to reduce energy consumption. These platforms can integrate with Building Management Systems (BMS) for centralized control of large commercial HVAC installations.

Integration with Smart Homes and Home Automation

An innovative feature of smart HVAC systems is their ability to integrate with other smart devices within a home or building. Advanced systems can communicate with automated blinds, smart lighting, and security systems, adjusting climate control based on overall environmental conditions.

For example, a smart HVAC system can work in synergy with temperature sensors in specific rooms, adjusting heating or cooling based on actual needs in each area. This zoned approach prevents energy waste and allows personalized comfort in different parts of the home or office.

Another innovation is the ability to receive automatic software updates, which enhance performance and introduce new features without requiring hardware replacement. This makes smart HVAC systems a flexible and future-proof solution.

The Future of Connected HVAC Systems

The HVAC industry is continuously evolving, and in the coming years, we will see the increasing adoption of more interconnected and automated systems. Artificial Intelligence (AI) and Machine Learning technologies are making HVAC systems even more efficient, with algorithms capable of predicting energy needs and adjusting climate control based on user behavior.

Another emerging innovation is the use of renewable energy to power HVAC systems, such as integration with solar panels and battery storage. This not only reduces dependence on traditional electrical grids but also allows for better energy utilization during peak hours.

Connected HVAC systems will play a crucial role in the development of smart cities, where buildings and infrastructure communicate with each other to optimize large-scale energy consumption. The adoption of these technologies will bring economic and environmental benefits, reducing the carbon footprint of traditional climate control systems.

9.2 Integration with Home Automation and Voice Assistants

The integration of HVAC systems with home automation and voice assistants represents one of the most significant advancements in climate control technology. Thanks to advanced connectivity, users can control heating, cooling, and indoor air quality simply by using voice commands or dedicated apps. This level of automation enhances home comfort while optimizing energy consumption, reducing operating costs, and minimizing environmental impact.

In the United States, the smart home market is continuously expanding, with more homes and offices adopting smart devices to improve efficiency and indoor security. Smart HVAC systems seamlessly integrate with platforms like Amazon Alexa, Google Assistant, and Apple HomeKit, providing users with unprecedented control over climate settings.

How HVAC and Home Automation Integration Works

The integration between HVAC and home automation is based on communication between different devices through network protocols such as Wi-Fi, Zigbee, and Z-Wave. The smart thermostat acts as a bridge between the HVAC system and the home automation system, allowing remote control via apps or voice assistants.

When an HVAC system is connected to a smart home network, it can exchange data with other devices such as temperature sensors, motorized dampers, and automated blinds. This enables climate adjustments based on real user needs, improving system efficiency.

For example:

- If a motion sensor detects that a room is unoccupied, the system can automatically reduce heating or cooling in that area, preventing energy waste.
- Similarly, an air quality sensor can activate ventilation when it detects increased levels of CO_2 or fine particulate matter.

HVAC Control via Voice Assistants

One of the most appreciated features of HVAC and home automation integration is the ability to control the system using voice commands. Leading voice assistants offer advanced climate management functions, making HVAC interaction more intuitive and accessible.

With Amazon Alexa, users can set specific temperatures with simple commands such as:
- *"Alexa, set the temperature to 72 degrees."*

Google Assistant offers similar capabilities, allowing users to create custom routines to automate HVAC operations based on time or weather conditions. A command like:
- *"Hey Google, turn on the heating at 6 AM."*

helps preheat the house before waking up, improving comfort while reducing energy consumption.

Apple HomeKit integrates HVAC control within its platform, enabling users to manage climate settings directly from the Home app on iPhone or iPad. With Siri shortcuts, advanced automation can be set up, such as:
- *"Siri, activate night mode."*

which automatically lowers the temperature for a more comfortable sleep.

Advanced Automations and Scheduling

A major advantage of HVAC and home automation integration is the ability to create advanced automations. Scheduled routines allow automatic climate regulation based on external factors such as weather conditions, time of day, or occupant presence.

For example:

- A smart HVAC system can be configured to reduce cooling during cooler evening hours and increase it during the day, using weather forecasts as a reference.

- When connected to a home weather station, the system can adjust air conditioning operation based on humidity and outdoor temperature, improving energy efficiency.
- Automations can also be linked to home security systems. When the alarm system is set to *"Away Mode"*, the smart thermostat can automatically switch to energy-saving mode, preventing unnecessary energy use.

Integration with Environmental Sensors and Smart Home Systems

Modern smart HVAC systems can work in synergy with various environmental sensors to enhance comfort and efficiency. Multi-zone temperature sensors allow heating or cooling adjustments based on the needs of each room, ensuring a more uniform climate distribution.

Air quality sensors monitor pollutant concentrations and activate ventilation when needed. This is particularly beneficial for homes with allergy sufferers or individuals with respiratory issues, as it ensures cleaner and healthier air.

Another valuable integration is with smart blinds and shades. If the HVAC system detects a rise in indoor temperature, it can communicate with motorized blinds to lower them, reducing heat gain and decreasing the air conditioner's workload.

The Future of HVAC and Home Automation Integration

The integration of HVAC and home automation is expected to evolve further in the coming years, with systems becoming increasingly intelligent and interconnected. Artificial Intelligence (AI) and Machine Learning will enable HVAC systems to predict user needs, analyzing behavioral patterns and automatically optimizing climate control.

Another emerging innovation is energy management through smart grids. Future HVAC systems will be able to communicate with electric grids to optimize consumption based on renewable energy availability, contributing to the development of low-impact homes and buildings.

As the adoption of HVAC-compatible smart home devices increases, manufacturers are working toward unified standards to enhance interoperability across different brands. This will make HVAC and smart home integration even more seamless and accessible.

9.3 Remote Monitoring and Predictive Maintenance

The HVAC industry is undergoing a significant transformation thanks to the introduction of remote monitoring and predictive maintenance. These technologies enable real-time system monitoring and help prevent failures before they occur, improving operational efficiency and reducing maintenance costs.

In the United States, where HVAC systems are essential for both home comfort and commercial building efficiency, the adoption of these solutions is rapidly growing. The ability to predict problems and intervene proactively is particularly useful for businesses, which can reduce downtime, and homeowners, who can avoid costly emergency repairs.

What Is Remote Monitoring and How Does It Work?

Remote monitoring in HVAC systems relies on a network of advanced sensors and IoT (Internet of Things) devices that collect real-time data on system performance. These data points are transmitted to centralized management software, which analyzes key parameters such as:

- Temperature fluctuations
- Refrigerant pressure levels
- Energy consumption
- Indoor air quality

Through a digital interface accessible via smartphones, tablets, or computers, users can monitor their HVAC system's status from anywhere. If an anomaly is detected, the system can send instant alerts, allowing for quick intervention before the issue worsens.

This technology is particularly beneficial for large commercial facilities, where monitoring multiple HVAC units can be complex and requires real-time responses to ensure comfortable environments and compliance with energy efficiency regulations.

Predictive Maintenance: Preventing Failures Before They Happen

Predictive maintenance is an evolution of traditional scheduled maintenance. Instead of conducting routine inspections based on a fixed schedule, this technology leverages AI algorithms to analyze HVAC operational data and predict potential failures.

For example:

- If a sensor detects an abnormal rise in compressor temperature or a drop in refrigerant pressure levels, the system can issue an alert, indicating that preventive action is needed before damage occurs.
- This approach reduces the risk of unexpected breakdowns and allows technicians to optimize maintenance schedules, avoiding unnecessary service visits and costly emergency repairs.

For industrial plants and large commercial buildings, predictive maintenance lowers energy consumption, extends equipment lifespan, and minimizes system downtime.

Benefits of Remote Monitoring for Technicians and Users

The implementation of remote monitoring and predictive maintenance offers significant advantages for both HVAC technicians and end users.

☑ For Technicians:

- Optimized time and resource management – interventions are performed only when necessary, and diagnostics are available before a site visit.
- Preemptive troubleshooting – technicians know the issue in advance and can bring the correct replacement parts, reducing repair time.
- Increased service efficiency – less frequent emergency calls, leading to better planning and cost control.

☑ For Homeowners & Businesses:

- Greater control over HVAC systems – users can track energy consumption patterns and optimize settings to lower utility bills.
- Instant failure detection – alerts are sent before an issue worsens, preventing long-term system damage.
- Performance tracking – historical data can be stored and analyzed to identify recurring inefficiencies and improve overall HVAC operation.

Tools and Technologies for Monitoring and Diagnostics

To implement remote monitoring and predictive maintenance, HVAC systems must integrate various tools and technologies. The most commonly used devices include:

- Temperature and humidity sensors – monitor indoor comfort and system efficiency.
- Digital pressure gauges – measure refrigerant pressure and detect leaks.
- Vibration analyzers – identify early signs of wear in mechanical components such as motors and compressors.
- Thermal imaging cameras – detect heat loss and verify proper airflow distribution in ducts.
- HVAC management software – collects sensor data, generates performance reports, and recommends maintenance actions.

Using these advanced tools helps technicians diagnose issues more accurately and minimize unexpected failures.

The Future of HVAC Monitoring: Automation and Artificial Intelligence

The rapid advancement of technology is pushing HVAC monitoring to an even more sophisticated level, thanks to automation and AI-driven systems. New HVAC models are increasingly equipped with machine learning algorithms capable of analyzing vast amounts of data and autonomously optimizing system performance.

✅ Key Future Trends:

- AI-powered HVAC control systems – adapting to user habits for maximum energy savings.
- Smart grid integration – HVAC systems will communicate with renewable energy networks to optimize power consumption.
- Digital twin technology – creating virtual models of HVAC systems to simulate performance, detect inefficiencies, and test improvements before implementing them in real-world applications.

With these innovations, remote monitoring and predictive maintenance will continue to redefine HVAC service efficiency, ensuring lower energy costs, improved environmental sustainability, and extended equipment lifespan.

9.4 HVAC and Sustainability: Solutions to Reduce Environmental Impact

The HVAC industry plays a significant role in energy consumption and environmental impact. Heating, ventilation, and air conditioning systems account for a substantial portion of global energy use and contribute to greenhouse gas emissions. For this reason, adopting sustainable solutions is essential to reduce environmental impact and improve building efficiency.

In the United States, the Department of Energy (DOE) and the Environmental Protection Agency (EPA) have introduced stricter regulations to promote the use of more efficient and eco-friendly HVAC systems. New technologies provide concrete opportunities to reduce energy consumption, minimize waste, and incorporate renewable energy sources, helping to transition toward a more sustainable future.

The Importance of Sustainability in the HVAC Industry

Traditional HVAC systems consume large amounts of energy for heating and cooling, often wasting resources due to outdated systems or inefficient energy use. This not only increases operational costs for homes and businesses but also significantly impacts CO_2 emissions.

One of the major concerns regarding HVAC systems is the use of refrigerants with high Global Warming Potential (GWP). Some previously common refrigerants, such as R-22, have been phased out due to their harmful effects on the ozone layer. Today, more environmentally friendly alternatives like R-32 and R-290 are becoming widespread, significantly reducing environmental impact.

Sustainable HVAC Technologies

To improve HVAC system sustainability, many companies are developing innovative technologies that reduce energy consumption and optimize efficiency.

✅ High-efficiency heat pumps are one of the most effective solutions to reduce fossil fuel usage for heating buildings.
✅ Geothermal HVAC systems use the stable underground temperature to heat and cool buildings with lower energy consumption. While initial installation costs can be high, the long-term benefits in terms of savings and sustainability are significant.

☑ Smart HVAC systems enhance efficiency through advanced temperature control and real-time consumption monitoring. Smart thermostats, for example, learn user habits and automatically adjust climate settings to prevent energy waste.

Renewable Energy and HVAC

One of the most effective strategies to reduce HVAC system environmental impact is integration with renewable energy sources. More buildings are adopting solar photovoltaic panels to power HVAC systems, reducing dependence on the traditional power grid and lowering CO_2 emissions.

In advanced systems, energy produced by solar panels is stored in battery storage units, allowing HVAC operation even at night or during low sunlight conditions. This approach is particularly beneficial in high-sunlight regions, such as the southern United States.

In addition to solar energy, another sustainable solution is assisted natural ventilation, which utilizes controlled airflow to reduce the workload of air conditioners and improve indoor air quality.

Regulations and Incentives for Green HVAC Systems

In the United States, the federal government and local authorities offer financial incentives to encourage the installation of high-efficiency HVAC systems. Programs like the Federal Tax Credit for Energy Efficiency provide tax deductions for replacing outdated systems with sustainable solutions.

The ENERGY STAR program, certified by the EPA, helps consumers identify high-efficiency HVAC systems. Choosing an ENERGY STAR-certified system can reduce energy consumption by up to 30% compared to traditional models.

Some states also offer specific incentives for using renewable technologies, such as grants for installing geothermal heat pumps or low-interest financing for HVAC systems connected to smart energy distribution grids.

Strategies for More Sustainable HVAC Usage

In addition to choosing high-efficiency HVAC systems, there are various strategies that users and technicians can adopt to improve system sustainability.

✅ Regular maintenance is essential to ensure maximum performance, preventing energy waste due to dirty filters, refrigerant leaks, or malfunctioning components.

✅ Zoning systems allow heating or cooling only occupied areas, reducing overall energy consumption. This approach is especially beneficial in large buildings, where not all rooms require the same climate control.

✅ Proper building insulation plays a crucial role in reducing energy demand. Good insulation prevents heat loss in winter and overheating in summer, minimizing HVAC system use.

9.5 How to Choose a Modern and Energy-Efficient HVAC System

Selecting a modern and efficient HVAC system is a decision that significantly impacts home comfort, energy consumption, and environmental footprint. With technological advancements and the introduction of stricter energy efficiency regulations, consumers in the United States now have many options to choose an HVAC system that delivers optimal performance with minimal energy waste.

A well-thought-out choice considers several factors, including system type, building size, thermal insulation quality, and integration with smart management systems. Understanding these elements helps select the most suitable solution, avoiding energy waste and ensuring long-term efficiency.

Evaluating the Building's Heating and Cooling Needs

One of the most common mistakes when choosing an HVAC system is selecting an oversized or undersized unit. An overpowered system will consume more energy than necessary, while an underpowered system will struggle to maintain a comfortable temperature during extreme weather conditions.

To determine the correct thermal load, it is essential to perform a Manual J Load Calculation, which considers factors such as home size, sun orientation, insulation quality, and window quantity.

Types of Energy-Efficient HVAC Systems

There are multiple options for modern and efficient HVAC systems.

☑ High-efficiency heat pumps are one of the best solutions for versatility and low energy consumption. These systems provide both cooling and heating while using electricity more efficiently than traditional gas or resistance heating systems.
☑ Zoned HVAC systems allow climate control only in needed rooms, avoiding energy waste. With smart thermostats and motorized dampers, air distribution is optimized based on each zone's needs.
☑ VRF (Variable Refrigerant Flow) systems are ideal for large buildings and commercial applications, offering precise temperature control by continuously adjusting the refrigerant flow.

The Importance of Energy Efficiency Ratings

In the United States, choosing an energy-efficient HVAC system involves evaluating certifications and energy efficiency labels.

☑ The ENERGY STAR program, certified by the EPA, ensures that an HVAC system meets strict energy-saving standards, balancing performance and consumption.
☑ SEER (Seasonal Energy Efficiency Ratio) measures air conditioner and heat pump cooling efficiency. Higher SEER values mean lower energy consumption.
☑ HSPF (Heating Seasonal Performance Factor) evaluates heating efficiency. The higher the HSPF, the more efficient the system.

Some states offer tax incentives for installing HVAC units with SEER ratings above a certain threshold, making these solutions even more cost-effective.

Smart Technologies to Maximize Efficiency

Integration with smart management systems is another key factor in choosing a modern HVAC system.

✅ Smart thermostats, like Google Nest or Ecobee, optimize energy consumption by automatically adjusting temperature settings based on occupant habits and weather conditions.
✅ Some advanced HVAC systems are compatible with smart grids, allowing them to adjust consumption based on peak energy demand periods, contributing to better energy distribution.
✅ Environmental sensors further improve efficiency by monitoring humidity levels, air quality, and temperature in each room, enabling optimized climate control.

Upfront Cost vs. Long-Term Savings

When choosing a new HVAC system, it is crucial to consider not only the initial cost but also long-term savings. A high-efficiency system may have a higher upfront price, but it ensures lower operating costs over the years by reducing energy consumption.

✅ Some manufacturers offer extended warranties on critical components like compressors and heat exchangers, making the investment more secure.
✅ Many local utilities in the U.S. provide rebates and discounts for replacing old systems with high-efficiency models, helping to lower the initial purchase cost.
✅ Proper installation is crucial. A poorly installed system can reduce efficiency by up to 30%, making it essential to hire certified HVAC technicians.

BOOK 10
Conclusion and Final Checklist

10.1 Summary of Key Concepts

After exploring the HVAC industry in detail—from its basic operations to advanced technologies—it's useful to recap key concepts to consolidate acquired knowledge. Understanding the role of each component, the importance of maintenance, and strategies to enhance energy efficiency allows users to make the most of an HVAC system, ensuring comfort, safety, and cost savings.

An HVAC system is a complex system that includes heating, ventilation, and air conditioning, all essential for residential and commercial buildings. The heart of the system is the compressor, which works together with the condenser, evaporator, and refrigerant to regulate air temperature efficiently. Each component plays a precise role, and proper maintenance is essential to ensure optimal long-term performance.

Modern HVAC systems integrate smart thermostats and automation technologies to improve temperature control and reduce energy consumption. The adoption of smart solutions such as remote monitoring and predictive maintenance helps prevent failures and optimize system efficiency. These advancements, combined with low-impact refrigerants and renewable energy solutions, contribute to making HVAC systems more sustainable and cost-effective over time.

How an HVAC System Works

The basic principle of HVAC operation is based on the refrigeration cycle, which cools or heats a space by transferring heat from one point to another. During this process:

- ✅ The compressor compresses the refrigerant.
- ✅ The condenser releases excess heat.
- ✅ The evaporator absorbs heat from indoor air, creating a cooler environment.

In heat pump systems, the cycle can be reversed, allowing the system to provide heating in winter. This technology is an energy-efficient alternative to traditional heating methods, as it reduces energy consumption.

Another essential factor is ventilation, which ensures air exchange and prevents humidity and mold issues. Air filters play a critical role in maintaining indoor air quality, removing dust, allergens, and pollutants.

The Importance of Maintenance and Diagnostics

An HVAC system requires regular maintenance to ensure optimal performance. Key maintenance tasks include:

- Cleaning and replacing filters to prevent dust accumulation and improve air quality.
- Checking refrigerant levels, compressors, and fans to avoid malfunctions and extend system lifespan.
- Using diagnostic tools such as multimeters, thermal cameras, and HVAC pressure gauges to detect hidden issues before they cause costly breakdowns.

With technological advancements, remote monitoring solutions now allow real-time notifications on system performance, enabling maintenance scheduling based on actual operational data.

Energy Efficiency and Choosing the Right System

With rising energy costs and new environmental regulations, choosing an efficient HVAC system is more important than ever.

- ENERGY STAR-certified systems ensure significant savings, reducing consumption by up to 30% compared to traditional models.

☑ SEER (Seasonal Energy Efficiency Ratio) helps determine cooling efficiency.

☑ HSPF (Heating Seasonal Performance Factor) measures heat pump efficiency in heating mode.

Integrating renewable energy sources like solar panels and geothermal heat pumps further improves HVAC sustainability. Additionally, home automation and smart thermostats allow users to personalize HVAC operations based on their needs, reducing unnecessary energy consumption.

Future Technologies and Smart HVAC

The HVAC industry is continuously evolving, with new technologies making systems more intelligent and adaptable.

☑ VRF (Variable Refrigerant Flow) systems dynamically adjust refrigerant flow, improving temperature distribution and reducing energy use.

☑ Artificial Intelligence (AI) in HVAC is advancing, with systems that learn user habits and automatically optimize operations to maximize efficiency.

☑ Connected HVAC systems that communicate with smart grids will help balance energy consumption based on renewable energy availability and demand levels.

The Role of HVAC Technicians and the Importance of Training

The HVAC industry requires advanced technical skills and continuous updates on new technologies and regulations.

☑ Certified technicians must understand thermodynamics, electronics, and HVAC mechanics, along with safety procedures for handling refrigerants and installing advanced systems.

☑ Certifications like EPA Section 608 Certification are mandatory for handling refrigerants legally and ensuring professional service.

☑ Specializing in smart HVAC, energy efficiency, and geothermal systems offers career growth opportunities and increases competitiveness in the job market.

10.2 Annual HVAC Maintenance Checklist

Annual HVAC system maintenance is essential to ensure efficiency, longevity, and reduced energy consumption. A well-maintained system operates more reliably and helps prevent costly and unexpected failures. In markets like the United States, where climate variations can be extreme, scheduled maintenance is critical to ensuring optimal comfort year-round.

A systematic maintenance approach allows technicians to identify minor issues before they become major, improving system safety and reliability. To conduct a thorough inspection, it is useful to follow a detailed checklist that covers all the main HVAC components, from cooling and heating units to air filters, ducts, and thermostats.

Air Filter Inspection and Cleaning

One of the most important tasks in HVAC maintenance is cleaning or replacing air filters. Dirty filters reduce airflow, causing system overload and higher energy consumption. On average, filters should be checked monthly and replaced at least every 90 days, but in dusty environments or homes with allergy sufferers, more frequent replacements may be necessary.

Clogged filters not only affect system efficiency but also reduce indoor air quality, leading to an accumulation of dust, allergens, and mold. A clean filter ensures proper airflow and prevents the HVAC system from working harder than necessary to maintain a comfortable temperature.

Inspection and Cleaning of Condenser and Evaporator Coils

Over time, condenser and evaporator coils accumulate dirt and debris, reducing their heat exchange efficiency. An annual check is crucial to prevent a decline in performance. The condenser coil, located in the outdoor unit, is particularly exposed to dust, leaves, and environmental debris.

Regular cleaning of these components helps maintain efficiency and prevents problems such as compressor overheating. Effective coil cleaning can be done using a dedicated coil cleaner and a low-pressure water spray to remove built-up dirt.

Refrigerant Level Check and Leak Detection

An insufficient refrigerant level can drastically reduce system efficiency and damage the compressor. During annual maintenance, it is essential to check refrigerant pressure and inspect for leaks in the system.

Using HVAC pressure gauges and electronic leak detectors, technicians can identify abnormalities in the refrigeration circuit. If a leak is detected, it must be fixed immediately to prevent efficiency loss and increased energy consumption.

Modern HVAC systems can even monitor refrigerant levels remotely, sending alerts to technicians in case of anomalies.

Thermostat and Control System Check

The thermostat plays a key role in HVAC efficiency, regulating temperature based on user settings. An annual check ensures that the device is working correctly and that settings are optimized for energy savings.

For smart thermostats, it's important to verify:

☑ Proper Wi-Fi connection and firmware updates.

☑ Calibration to prevent discrepancies between set temperature and actual temperature.

☑ Automation settings to align HVAC operation with occupant habits for better efficiency.

Inspection of Fans, Motors, and Drive Belts

HVAC system fans and motors must be inspected to ensure proper operation and detect any abnormal vibrations. Drive belts, if present, should be checked for signs of wear or looseness.

A malfunctioning fan can reduce airflow and increase system strain, leading to higher energy use and potential mechanical failures. Regular lubrication of fan bearings and replacement of worn-out belts helps prevent unexpected cooling or heating issues.

Advanced HVAC systems now feature vibration sensors that detect irregular patterns in fan operation and alert technicians before a major failure occurs.

10.3 Essential Tools for Every HVAC Technician

The success of an HVAC technician depends not only on acquired skills but also on the quality and adequacy of the tools used. A well-equipped kit allows for accurate diagnostics, efficient repairs, and installations that meet safety and energy efficiency standards.

In the United States, the HVAC industry follows strict regulations and requires certified tools to ensure precise and safe work. Technicians must be equipped with:

✓ Basic hand tools
✓ Advanced measuring devices
✓ Specialized equipment for refrigerant handling and system performance analysis

Technological advancements have greatly improved diagnostic precision, making digital tools and remote monitoring essential. A well-prepared HVAC professional must select the most suitable tools for the job, avoiding obsolete or non-compliant equipment. Investing in high-quality tools can make the difference between a quick and precise intervention and a lengthy, inaccurate job that could compromise system reliability.

Hand Tools and Basic Equipment

Every HVAC technician must have a set of essential hand tools for installations, repairs, and routine maintenance:

☑ Adjustable wrenches, pliers, screwdrivers, and tube cutters – essential for mechanical components and electrical connections

☑ Clamp meters – allow current measurement without disrupting the circuit

☑ Insulated screwdrivers – ensure safety when working on electrical panels

☑ Hex keys (Allen wrenches) and spanners – needed for loosening and tightening HVAC system fittings

☑ Tube cutters – used to precisely cut copper and aluminum pipes, preventing refrigerant leaks from faulty connections

✓ A clean, precise cut ensures better joint sealing and extends system lifespan.

Electrical Diagnostic Tools

Many HVAC issues stem from electrical failures, so technicians must use specific tools to test and analyze the system's electrical components:

✓ Digital multimeter – measures voltage, current, and resistance, helping detect broken circuits or faulty components

✓ Non-contact voltage detector – quickly identifies live wires, reducing the risk of electric shock

✓ Portable oscilloscope – useful for analyzing electrical signals, identifying control circuit issues in modern HVAC systems

Modern HVAC systems often have electronic control boards and advanced sensors, making electrical diagnostics a critical step in system troubleshooting.

Refrigerant Measurement Tools

Checking the refrigerant level and system pressure is one of the most critical tasks to ensure proper HVAC system performance:

✓ Digital HVAC manifold gauge – accurately measures refrigerant pressure, identifying leaks or load imbalances

✓ Refrigerant leak detector – crucial for detecting gas leaks in refrigeration circuits

✓ Vacuum pump – removes air and moisture from the system before charging refrigerant, ensuring a contaminant-free refrigeration circuit

! Refrigerant leaks reduce system efficiency, pose environmental risks, and may violate federal refrigerant regulations.

Airflow and Indoor Air Quality Monitoring Tools

In modern HVAC systems, maintaining indoor air quality (IAQ) is essential for occupant comfort and health.

✓ Digital anemometer – measures air velocity in ducts to ensure proper distribution

✓ Hygrometer – monitors indoor humidity levels; excessive humidity promotes mold and bacteria, while low humidity can cause respiratory discomfort

✓ CO₂ and particulate sensor – detects ventilation issues and monitors air pollutants

✓ These tools help verify that the HVAC system is operating at optimal efficiency and ensuring clean air quality.

Specialized Tools and Advanced Technologies

Technological advancements have introduced high-precision tools that enhance HVAC diagnostics:

✓ Infrared thermal cameras – essential for detecting thermal leaks, insulation failures, and critical points in air distribution
✓ Vibration sensors – monitor motors and fans, preventing mechanical failures before they become serious issues
✓ Remote monitoring software – collects real-time system performance data, facilitating troubleshooting from a distance

HVAC smart systems increasingly rely on data-driven insights, allowing technicians to implement predictive maintenance and improve system reliability.

10.4 Resources and Recommended Readings

The HVAC industry is constantly evolving, with new technologies, regulations, and work methodologies emerging regularly. To stay up to date and improve skills, it is crucial to access reliable information sources and deepen knowledge through technical manuals, training courses, and online platforms.

In the United States, numerous resources are available for HVAC professionals, from beginners to experienced technicians. In addition to reference books, there are associations, websites, and training channels offering updated content, webinars, and certifications. Using these resources helps enhance technical expertise, stay informed about latest regulations, and adopt best practices for installation, maintenance, and repair.

Essential HVAC Books for Technicians

For those looking to deepen their understanding of HVAC systems, several technical manuals are considered indispensable in the industry:

✓ Modern Refrigeration and Air Conditioning – a comprehensive guide covering refrigeration, air conditioning, and ventilation systems. Frequently used in HVAC training courses, it provides detailed explanations of basic theories, installation practices, and maintenance procedures.

✓ HVAC Systems Design Handbook – offers an overview of HVAC system design strategies and selection criteria for residential and commercial buildings. This book is especially useful for those specializing in high-efficiency HVAC system design.

✓ Troubleshooting HVAC-R Systems – a must-have guide for diagnosing and resolving common HVAC system issues. This book includes real-world case studies and practical troubleshooting tips to quickly identify and fix problems.

Training Courses and Certifications

Beyond books, attending training courses is a crucial step for acquiring and validating HVAC skills. In the United States, many technical schools and professional associations offer specialized HVAC programs.

✓ EPA Section 608 Certification – one of the most essential courses, required for handling refrigerants. This certification ensures that a technician has the necessary knowledge to safely manage refrigerants while complying with environmental regulations.

✓ NATE (North American Technician Excellence) – a highly regarded certification that validates advanced HVAC skills in installation, maintenance, and repair. Earning this certification enhances career opportunities and demonstrates a high level of professional expertise.

✓ BPI (Building Performance Institute) Certification – focuses on energy efficiency optimization in HVAC systems, helping professionals specialize in sustainable solutions for reducing energy consumption in buildings.

Websites and Online Resources

The internet provides a vast amount of HVAC information, including technical articles, practical guides, and industry updates. Some top websites include:

✓ Air Conditioning Contractors of America (ACCA) – a leading industry resource providing installation standards, regulations, and training courses.

✓ HVAC-Talk – an online community where HVAC technicians share experiences, solve problems, and discuss the latest innovations.

✓ HVAC School YouTube Channel – features detailed video lessons covering everything from common HVAC issues to advanced maintenance techniques.

✓ Energy Star – offers important insights on energy efficiency requirements for HVAC systems, helping technicians and consumers select the best solutions available.

Apps and Software for HVAC Professionals

Technological advancements have led to the development of many mobile apps that help HVAC technicians work more efficiently. Some of the most useful HVAC apps include:

✓ HVAC Buddy – an excellent tool for checking refrigerant charge levels and diagnosing overcharging or undercharging issues.

✓ MeasureQuick – collects real-time system data from HVAC sensors, helping technicians identify anomalies and optimize performance.

✓ Elite Software Rhvac – a leading software for load calculation and HVAC system component selection, widely used for HVAC design.

HVAC Industry Events and Conferences

Attending events and conferences is an excellent way to stay up to date on the latest industry innovations and network with other professionals.

✓ AHR Expo – one of the largest HVAC industry trade shows in the U.S., bringing together manufacturers, engineers, and technicians from around the world.

✓ National HVACR Education Conference – an unmissable event for HVAC professionals, offering workshops and seminars conducted by industry experts.

✓ ASHRAE (American Society of Heating, Refrigerating, and Air-Conditioning Engineers) – hosts events and webinars on energy efficiency and technological innovation in HVAC.

10.5 Next Steps: How to Continue Advancing in the HVAC Industry

Knowledge in the HVAC sector does not end with this book. The industry is constantly evolving, with new technologies, stricter regulations, and increasing demands for energy efficiency and sustainability.

For this reason, anyone who wants to stand out as an HVAC technician or deepen their understanding of HVAC systems must adopt a continuous learning approach.

For many HVAC professionals, the learning journey never truly ends—new installation methods, advanced diagnostic tools, and innovative efficiency solutions emerge every day. Staying up to date is not just a necessity for fieldwork but also an opportunity to increase market value and access higher-paying specializations.

Specializing in a Specific HVAC Field

After mastering basic skills, it is beneficial to consider specialization. HVAC encompasses various sectors, each with unique technical aspects and career opportunities.

✓ Commercial and Industrial Refrigeration Systems – These systems are widely used in supermarkets, hospitals, food industries, and restaurants. Working in this field requires in-depth knowledge of refrigerant management, thermal balancing, and environmental regulations.

✓ High-Efficiency HVAC Systems – This rapidly growing sector focuses on geothermal heat pumps, VRF systems, and renewable energy integration. Qualified technicians for these advanced systems are in high demand, especially with government incentives for reducing CO_2 emissions.

✓ Indoor Air Quality (IAQ) – An increasingly crucial specialization in both residential and commercial applications. This area includes advanced ventilation system design, indoor pollutant monitoring, and humidity optimization to ensure healthy indoor environments.

Earning Advanced Certifications and Professional Qualifications

To stand out in the HVAC industry, obtaining nationally recognized certifications is essential. The EPA Section 608 Certification, mandatory for handling refrigerants, is just the first step. Those aiming for higher-level positions should pursue additional certifications that validate specialized skills.

✓ NATE (North American Technician Excellence) – One of the most prestigious certifications in the U.S., recognized by employers and HVAC manufacturers. It demonstrates advanced skills in installation, maintenance, and system diagnostics.

✓ HVAC Excellence Certification – Offers specialized programs in refrigeration, electronic diagnostics, and high-efficiency HVAC system design. Holding these credentials can lead to higher-paying jobs and supervisory or training roles.

Using Technology to Improve Skills

Today, HVAC training is no longer limited to traditional courses. With technology, professionals can access e-learning platforms, virtual simulators, and interactive webinars to improve HVAC knowledge without needing to attend in-person classes.

✓ Virtual Training Software – Some simulation programs allow HVAC technicians to practice troubleshooting and manage complex HVAC systems in a virtual environment, improving operational understanding without risking real equipment damage.

✓ HVAC Mobile Apps – The use of HVAC apps is growing. Apps like MeasureQuick, HVAC Buddy, and iManifold provide advanced diagnostic tools and step-by-step troubleshooting guides to help technicians identify issues and optimize system performance.

✓ Online HVAC Communities – Engaging in technical forums like HVAC-Talk allows technicians to connect with industry experts, exchange experiences, and receive advice from seasoned professionals.

Attending HVAC Events, Conferences, and Trade Shows

Industry events and conferences are excellent opportunities to stay updated on the latest HVAC innovations, discover new tools, and network with other professionals.

✓ AHR Expo – One of the largest HVAC trade shows, where companies and professionals worldwide showcase new products and cutting-edge solutions. Attending this event allows technicians to test new equipment, watch live demonstrations, and interact with top HVAC manufacturers.

✓ ASHRAE Conferences – Hosted by the American Society of Heating, Refrigerating, and Air-Conditioning Engineers, these conferences offer technical workshops and seminars on new regulations, energy-saving solutions, and best installation and maintenance practices.

Building a Strong Career and Looking to the Future

For HVAC professionals, the next step may involve starting a business or further specializing to advance into higher-level roles, such as HVAC supervisor or technical instructor.

✓ Becoming an HVAC Contractor – For those wanting to start their own business, obtaining an HVAC contractor license is required in many U.S. states. Running a business requires technical expertise as well as knowledge of business management, cost planning, and marketing strategies to attract clients.

✓ Advancing to Higher-Level Roles – HVAC technicians can also progress to lead technician or project manager positions, overseeing large system installations or managing teams on commercial and industrial projects.

✓ The Future of HVAC Careers – The HVAC industry is expected to grow significantly, with a strong emphasis on energy efficiency and sustainability. Technicians with expertise in smart HVAC systems, heat pumps, and advanced refrigeration technologies will have the best career opportunities.

Made in the USA
Las Vegas, NV
31 May 2025